AIRBUS
A300

AIRLIFE'S AIRLINERS 8

AIRBUS A300

Günter Endres

Airlife
England

First published in the UK in 1999
by Airlife Publishing Ltd

British Library Cataloguing-in-Publication Data
A catalogue record for this book
is available from the British Library

ISBN 1 84037 069 6

Printed in Singapore by Kyodo Printing Co (S'pore) Pte Ltd

Airlife Publishing Ltd
101 Longden Road, Shrewsbury, SY3 9EB, England
E-mail: airlife@airlifebooks.com
Website: www.airlifebooks.com

COVER: Monarch Airlines A300B4-605R.

PREVIOUS PAGE: Lufthansa A300B4-603.

BELOW: Laker Airways A300B4-203.

Acknowledgements
It is true to say that without the excellent co-operation from
Airbus Industrie, I would have struggled to complete this
enjoyable but enormous task. My sincere gratitude go to
Barbara Kracht, General Manager Press and Information
Service, and David Velupillai, Regional Manager Press
Relations, for arranging my visit to Toulouse and persuading
senior managers to take time out to talk to me. I would particu-
larly like to thank Bernard Ziegler, Airbus' former Senior Vice-
President Engineering, who gave up his precious leisure time to
give me his unique insight into the development of Airbus, to
David Lenormand, Manager Product Marketing, Arnaud
Martin, Director Transport Operations, and the staff in the
photographic archives.

As always, I have been fortunate to be able to call on many
long-standing friends and colleagues, who all gave their time
freely to fill gaps and make necessary corrections. Ricky-Dene
Halliday again kindly supplied some of the photographs from
his extensive collection, and the airlines themselves, though not
nearly enough, did their part in making this book as compre-
hensive as is reasonably possible. David Carter, who with
Graham Humberstone checked for technical accuracy in the
previously-published DC-10 book, again did a similarly thor-
ough job on the A300. To all, my sincere appreciation.

Günter Endres
West Sussex

CONTENTS

1 BACKGROUND AND EVOLUTION

In retrospect, the decision to develop a pan-European aircraft industry to challenge the domination of US manufacturers — a leap of faith both politically and economically — has proved to have been right. Without the benefit of a long history, Airbus Industrie has today achieved a worldwide market share of more than 40 percent, and is well on course to achieving parity with the mighty Boeing Airplane Company as the new millennium begins. But in the early 1960s, when the concept of a new European-wide aircraft was first mooted, there were many who expressed serious doubts about this venture, as well as those who simply ridiculed the idea, especially on the other side of the Atlantic. For three countries, with a recent history of conflict still fresh in the mind, to work together on developing an unproven concept of a large twin-engined airliner, seemed fraught with dangers. Yet, the fear of delays, disagreements and indecision, proved largely unfounded, and even when dithering Britain eventually pulled out, Germany and France had enough determination to succeed.

While politically and economically, European collaboration as envisaged was faced with many, sometimes seemingly insurmountable hurdles, technologically, Europe could not only hold its own, but often led the world in radical new thinking. Where it failed dismally, was in its inability to fully exploit its advances, while the US made full use of quantity production and proceeded to capture a major part of the civil market. In the first two decades after World War 2, only Britain and France were able to make a significant contribution to the development of civil aviation, but even so, the Vickers Viscount and the Sud-Aviation Caravelle were the only aircraft sold in appreciable numbers on the world market.

Interestingly, the Caravelle represented a first tentative attempt at European collaboration, utilising the nose and flight-deck of the de Havilland Comet, which led to a more substantial and wide-ranging partnership between the French and British industries on the supersonic Concorde. Unfortunately, Concorde, while a tremendous technological achievement and still the only supersonic aircraft in service, swallowed up most of the available research and development money, with the result that the Americans stole a march on Europe in the subsonic medium- to long-haul jet market. But the emerging mass travel era, made possible by powerful new jet engines and plunging air fares, directed attention to high-capacity short-

BELOW LEFT AND ABOVE: The Caravelle (left) — seen at Newcastle Airport in 1992 in Air Toulouse colours — and the Viscount (above) — a 1960s view of a Viscount 800 at Gatwick — were the only European designs to be sold in appreciable service on the world market. *Leo Marriott*

RIGHT: The 204-seat Hawker Siddeley HS 134 was one of several British designs of the 1960s.

range aircraft, with most manufacturers on both sides of the Atlantic beginning to develop a variety of interesting 'wide-body' concepts.

In Europe, these were loosely based on specifications and requirements developed in the UK by the Lighthill Committee and by a working party set up by the major European airlines. These appeared to narrow down to an aircraft between 200 and 250 seats, with a design range of around 800 nautical miles (1,500km). The description 'European Airbus' was already being suggested for this new type of aircraft. In the UK, Hawker Siddeley Aviation (HSA) was working towards high-capacity successors to the Trident, proposing the 160-seat plus HS.132 and the 204-seat HS.134. Both were to be powered by two new-technology rear-mounted 133.5kN (30,000lb) thrust Rolls-Royce RB178s. Across the Channel, Bréguet proposed the similar-sized, but double-deck Br124, powered by four Rolls-Royce Speys mounted in pairs under the wing. Alternative twin-engined layouts were also drawn up, using either the RB178 or Pratt & Whitney's JT9D turbofans. Nord Aviation, which had

already agreed to co-operate with Bréguet, had the N600 on the drawing board — a high-wing design, also with four Speys, but with an unusual horizontal double-bubble fuselage and 12-abreast seating for 250 passengers. The two side-by-side cabins were separated by a central bulkhead.

Sud-Aviation was working on the Galion, a single-deck version for 200 passengers, six-abreast, and a double-deck alternative for up to 250 passengers. France's other aircraft manufacturer, Avions Marcel Dassault, evaluated a 220-seat double-deck design with the engines mounted under a low wing. All manufacturers considered other imaginative and sometimes fanciful design concepts, including jet flaps, canards, tandem wings and many forms of different wing planform configurations.

LEFT: Signing of the Franco-German agreement for the joint development of the Airbus at Le Bourget on 29 May 1969 inside the mock-up. Britain had withdrawn a month earlier.

BELOW: The first artist's impression of the proposed twin-engined, widebody Airbus.

BELOW RIGHT: The Anglo-French HBN 100 design formed the basis of what was to become the Airbus.

BELOW FAR RIGHT: Full-scale mock-up of the A300B cockpit section at the Paris Air Show in 1969 adorned with the names of the participating manufacturers and their nations' flags.

TRILATERAL DISCUSSIONS

The governments of France, Germany and the United Kingdom quickly realised the enormous financial and technical investment that would be required for such an aircraft and urged their manufacturers to enter into partnerships. The British government in particular was leaning strongly towards European collaboration and in early 1964 drew up, jointly with the French, a paper entitled *An Outline Requirement for an Ultra High-Capacity Short-Range Aircraft*, which set out the guidelines for this new-generation of airliner, conditional upon cross-Channel collaboration. It is rather ironic that the British government later became the least committed party to the Airbus project.

BAC and Sud-Aviation, already working together on Concorde, met in July 1964 to discuss a 180-200-seat short/medium-range aircraft, but serious project discussions between France and the United Kingdom did not begin until June the following year. In that same month, France also initiated high-level talks with the German industry and, coincidentally, the subject also came up when the British Transport Minister visited Bonn. Germany saw collaboration as essential if its industry, which had been hampered by the restrictions imposed at the end of the war, was to get back into the mainstream of aircraft production. Its enthusiasm was evidenced by the Federal government's willingness to take a 25 percent stake in any future international venture, and the setting up the fol-

lowing month by Dornier, Hamburger Flugzeugbau (HFB), Messerschmitt, Siebelwerke and Vereinigte Flugtechnische Werke (VFW) of an Airbus Study Group at Munich.

The problem for Germany was how to create an organisational structure that allowed participation by the five companies, while ensuring a single interface with the international partners. This question was being addressed by a joint venture project team (Arbeitsgemeinschaft Airbus), which also set out to work towards presenting a joint German aircraft proposal. The joint project team started work in January 1966. Prompted by the UK government's rejection of the HS.134, HSA entered into discussions with Bréguet and Nord Aviation, while the British Aircraft Corporation initiated talks with Sud-Aviation and Dassault about possible co-operation on the Galion. The more fanciful ideas were discarded and what emerged from the Hawker/Bréguet/Nord team was a fairly conventional 200-250 seat aircraft.

Designated the HBN 100, it proposed a circular fuselage and two high by-pass ratio turbofan engines mounted under the low wing with a 30° sweep. The freight holds were located under the cabin floor. HBN also submitted four back-up designs, all of which were discarded. These included the high-wing HBN 101, based on the Nord N600 with a horizontal double-bubble fuselage; the HBN 102 and HBN 103 two-deck proposals with engines mounted under a mid-wing, but varying locations of freight holds; and the HBN 104, which was again a vertical two-deck design but adopted a clean low wing with the engines mounted on the rear fuselage. The two-deck designs were considered too problematic in terms of operations at airports, which were not then equipped for loading and unloading from different levels, and in the event of a ditching, where the safe evacuation of passengers from the lower deck was causing concern. These difficulties did not apply in the case of the HBN 101, but it too was considered too radical, leaving the HBN 100 as the design, which was to form the basis of what eventually became the Airbus.

AIRBUS ADVANCE

Still to be resolved were disagreements among airlines and manufacturers on the specific design requirements, particularly concerning capacity and range. This was discussed at the first intergovernmental meeting between Germany, France and Great Britain in March 1966, but little progress was made towards a definitive project definition until the aircraft manufacturers presented a joint concept document to the governments on 15 October 1966. The document was approved on 9 May 1967 at a trilateral ministerial meeting held in Paris, where the industry was also requested to supply a joint study based on two Rolls-Royce RB207 engines. This was submitted on 30 June, and on 25 July 1967, the three governments decided to proceed with the definition phase. It was then that the aircraft was first referred to as the A300, the designation indicating the approximate seating capacity of the aircraft. It had also grown in size to accommodate 267 passengers in a circular 21ft (6.4m) fuselage, and was powered by two RB207 engines generating a static thrust of 47,500lb (211kN) each.

In the meantime, with talks on the Galion project going nowhere and British European Airways (BEA) concerned that the Airbus would be too late to meet its requirements for a larger complementary type to the 1-11, BAC firmed up its 2-11 design. The BAC 2-11 was based on carrying 208 passengers a distance of 1,300 nautical miles (2,400km), and was claimed to be the quietest possible aircraft. BEA initially envisaged a requirement for 12 aircraft for its high-density business and tourist routes by summer 1972, but soon revised its estimate upward to between 30 and 40 aircraft. However the British government had committed itself to the Airbus project and, wanting to be seen as pro-European, refused to fund the 2-11.

This decision flew in the face of evidence gathered from European airlines by BAC, which pointed to a market for 1,500 for an aircraft of that size. There were many on the British side who questioned the government's logic of supporting a 300-seater, which nobody apparently wanted. If the 2-11 had one drawback, it was the fact that it was not a wide-body design.

The British decision came soon after the three governments had signed a Memorandum of Understanding (MoU) on 26 September 1967 authorising continued design studies and project definition. The agreement nominated Sud-Aviation, Hawker Siddeley and Deutsche Airbus, which had been founded earlier that month as a grouping of the five German manufacturers, as the airframe partners. After discarding earlier plans for total French leadership, with the others acting as subcontractors, it was also agreed that Sud-Aviation would have design leadership for the airframe, in exchange for Rolls-Royce leadership on the engine. Britain and France were to contribute 37.5 percent to the first phase development costs, with Germany providing the balance. Britain was also to be responsible for 75 percent of the engine development costs, while France and Germany shared the remainder. Rolls-Royce was to work with French engine manufacturer Snecma and MTU in Germany. The agreement also stipulated that construction of a prototype would only be authorised upon receipt of 75 firm orders from the three flag-carriers. A deadline of 31 July 1968 was set, giving 10 months to generate sufficient sales.

This was the real start of joint studies into an A300 Airbus, based on earlier work done on the HBN 100 and Galion projects, and the intense activity which followed led to project definition by June 1968. Broad agreement was also reached on the distribution of work, which roughly equalled the respective financial input of the partner nations. In addition to design leadership, Sud-Aviation took responsibility for the flightdeck, nose and fuselage centre section, engine installation, most of the systems definition and final assembly. Hawker Siddeley was given design and production responsibility for the wing, while Deutsche Airbus was charged with producing the remainder of the fuselage, empennage development, as well as

ABOVE: Air France was the driving force behind shaping the Airbus specification.

ABOVE RIGHT: The first contract for the Airbus is signed by Air France in November 1971.

definition and design of the passenger cabin, cargo holds and installation of the auxiliary power unit (APU).

But it certainly was not plain sailing from then on. In a moment of nationalistic fervour, the French government decided to cancel the Airbus project in favour of the Dassault Mercure, before reversing its decision three days later, and Rolls-Royce was changing direction towards the new US tri-jets, which it considered a more lucrative market for its smaller RB211 engine. The Airbus had also grown into a much larger aircraft, making increased thrust demands on the RB207, which was eventually dropped. Of the major European airlines, only Air France and Air Inter supported a 300-seat aircraft, with the others, especially Germany's flag-carrier Lufthansa, convinced that such a size aircraft would not be needed for another 10 years. It was also felt that the Lockheed TriStar and Douglas DC-10 under development in the United States, while intended primarily for longer stage-lengths, would, nevertheless, be capable of encroaching successfully on the shorter-range market envisaged for the European Airbus.

The lukewarm interest shown by the airlines forced the ministers of France, Britain and Germany to extend the design study for a further four months, when they met in Paris on 2 August 1968. The final design proposals were submitted at the end of October, but still failed to meet requirements and a final decision to go ahead was postponed once again, plunging the whole project in serious jeopardy. The cancellation of the Airbus became a real possibility and was not helped by Sud-

the three governments in October 1970, even though neither Air France nor Lufthansa had committed to buy the aircraft at that time.

Britain's withdrawal, nevertheless, presented France and Germany with a problem. While they were happy to proceed on their own, in spite of the increased financial commitment, neither country had the capacity, nor the facilities to take over the design and manufacture of the wing. Both eventually concluded that to avoid unacceptable delays to the programme, and to have too many airlines commit themselves to the new American designs, the retention of Hawker Siddeley was essential to move the project forward. It was to its great credit, that the British company decided to continue as a privileged subcontractor, partly using its own funds. A contract giving Hawker Siddeley responsibility for the design and construction of the complete wing box, as well as a share of sales and support activities, was signed during the Paris Air Show.

Aviation working on independent designs and pursuing tie-ups with other European manufacturers. In Britain, BEA showed a strong preference for the Rolls-Royce RB211-powered BAC 3-11, which was designed to accommodate 220 passengers over a distance of 1,740 nautical miles (3,220km).

In the face of overwhelming evidence for a smaller aircraft, details were released on 11 December 1968 by Hawker Siddeley and Sud-Aviation of a scaled-down version for 250 seats, designated at first the A250 and then the A300B. This brought the new Airbus within the ambit of the 50,000lb (222.5kN) thrust of the General Electric CF6-50 which was being proposed for the DC-10, the Pratt & Whitney JT9D, and the Rolls-Royce RB211 being developed for the Lockheed TriStar. The partners stated that the smaller size, together with a virtually existing engine, permitted a reduction in launching costs by approximately 30 percent.

While the airlines reacted favourably to the new definition, Britain threw a spanner in the works. Having three years earlier enthusiastically backed a 250-seat Airbus with Rolls-Royce engines, the new option of alternative US engines was used as a pretext for re-examining its position. In the absence of any order for the Airbus, it was already having misgivings about its £60 million investment and welcomed this ready-made excuse with open arms. After protracted negotiations at government level, it was announced on 10 April 1969 that Britain had withdrawn from the Airbus project. France and Germany, however, forged ahead, and on 29 May 1969 — the opening day of the 28th Paris Air Show — the German Minister of Economics Karl Schiller and the French Transport Minister M Jean Chamont signed the final development contract, which covered the prototype phase and extended over a period ending one year after certification of the basic type. A contract to move the Airbus into the construction phase was agreed between

STRUCTURING THE ORGANISATION

Following the withdrawal of the British government and Rolls-Royce, the major responsibilities for the Airbus A300 lay with the French and German authorities and, on the industrial side, with Sud-Aviation as design leader in Toulouse and Deutsche Airbus in Munich. Hawker Siddeley's role was smaller, but the British company remained very much involved in the evolution of the project. The organisation was rationalised and restructured on 18 December 1970, when Airbus Industrie was set up as a Groupement d'Intérets Economique (GIE) under French law. The legal framework for this form of corporation was set in an Ordinance of the French Republic, dated 23 September 1967. It was created specifically to facilitate the formation of co-operative ventures, for which existing legislation proved too restrictive. The GIE provides considerable operational flexibility and enables third parties to deal with a single entity. All members act as mutual guarantors for all commitments of Airbus Industrie third parties, and the legal structure, backed by government guarantees, ensures adequate cover to meet these commitments. The formation of the GIE as a central authority for technical, financial and marketing support, silenced some of the critics which had rubbished Europe's collective attempt as little more than an 'aircraft built by a committee', with no one taking responsibility for their actions.

In the interim a number of changes had taken place, some

LEFT: Henri Ziegler signs the loan agreement with the European Investment Bank (EIB) in Luxembourg in December 1971.

RIGHT: Two of the founding fathers of Airbus — Henri Ziegler. (LEFT) and Roger Beteille (RIGHT).

of which affected the composition of the GIE. In France, Sud-Aviation and Nord Aviation had merged on 1 January 1970 to form Aerospatiale, while in Germany, Dornier withdrew from Deutsche Airbus to pursue an independent path. Deutsche Airbus had been established on 4 September 1967, back-dated to 25 July, with a capital stock of DM5 million, each of the five companies — Dornier, HFB, Messerschmitt, Siebel and VFW — contributing DM1 million. A subsequent rationalisation of Germany's manufacturing capabilities brought together HFB, Messerschmitt and Siebel under the MBB (Messerschmitt-Bölkow-Blohm) banner, with VFW linking up with Fokker in the Netherlands. As a result, MBB undertook to contribute DM32.5 million, or 65 percent, of the new capital stock of DM50 million, with VFW providing DM17.5 million, or 35%.

At the end of 1970 the Dutch government decided to join the programme and an appropriate intergovernmental agreement was signed on 28 December, but Fokker-VFW, the company designated to be the Dutch participant, elected not to become a full member of Airbus Industrie. One year later, Spain came into Airbus, taking a 4.2% stake in equal portions from the German and French participation. The share was based on the value of the horizontal tailplane, which was to be taken over from MBB by Spanish manufacturer CASA. Shareholding in Airbus Industrie was then distributed among Aerospatiale (47.9%), Deutsche Airbus (47.9%) and CASA (4.2%). It remained so until Britain rejoined the consortium on 1 January 1979, taking a 20% stake, with 10% each coming from Aerospatiale and Deutsche Airbus.

FOUNDING FATHERS

The early progress, as well as the later success of the Airbus consortium, was shaped by just a handful of people from the three nations, all of whom played a crucial role in nursing the budding European alliance through infancy and into maturity, often with odds heavily stacked against them. All were inspira-tional characters and brilliant in their execution of a common vision for a healthy and effective European aircraft manufactur-ing industry, able to meet future requirements and to claw back some of the market share which had disappeared across the Atlantic.

Among the men who merit the accolade of founding fathers of Airbus were Roger Beteille, an aeronautical engineer who apportioned the work between the partners and kept a steady ship until his retirement as managing director in 1985, and General Henri Ziegler, first managing director of Airbus. One time French Resistance fighter, then president of Air France and later head of Bréguet and Sud-Aviation, it was Ziegler who had long ago realised that France, or any other European coun-try on its own, could not possibly hope to compete effectively with the aerospace budgets enjoyed by the United States manu-facturers. While this pan-European viewpoint sometimes caused consternation among the French political establishment, he was soon proved right. His death in July 1998 marked the end of an era, but his legacy remains in good hands.

The outstanding drivers on the German side were the late Franz-Josef Strauss, the ebullient and often controversial Bavarian politician, who was working tirelessly to re-establish the German aerospace industry following the ravages of war and became the first chairman of the Airbus Supervisory Board, and aerospace engineer Felix Kracht, who mastermind-ed the complicated and far flung network of manufacturing sites, which today still produce Airbus parts for final assembly in Toulouse and Hamburg. The contribution of Sir Arnold Hall, chairman of Hawker Siddeley, who exhibited exceptional courage in staying with Airbus when Britain pulled out and effectively kept the door open for Britain to rejoin later, should also not be underestimated, nor forgotten.

There were many other 'true Europeans' without whose team spirit and single-minded determination the Airbus dream could not have been turned into reality, against all the odds.

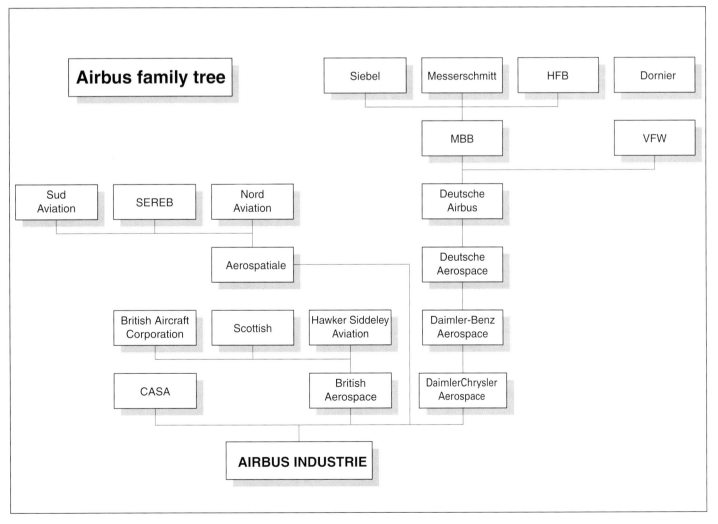

Airbus family tree

Siebel — Messerschmitt — HFB — Dornier

MBB — VFW

Sud Aviation — SEREB — Nord Aviation — Deutsche Airbus

Aerospatiale — Deutsche Aerospace

British Aircraft Corporation — Scottish — Hawker Siddeley Aviation — Daimler-Benz Aerospace

CASA — British Aerospace — DaimlerChrysler Aerospace

AIRBUS INDUSTRIE

2 DESIGN AND PRODUCTION

At the beginning of 1969, the outline design and overall defini-tion could be considered complete, and construction drawings were being finalised. Airline preferences had brought about a scaled-down model with a length of 167ft 2in (50.95m) and 18ft 6in (5.64m) fuselage diameter to accommodate 250 passen-gers. The fuselage cross-section had been subject to intensive investigations and was eventually fixed in size and shape to pro-vide comfortable eight-abreast seating in scheduled service, and to permit the accommodation of standard freight containers in the lower cargo hold. Economic considerations dictated a cruis-ing speed of up to Mach 0.84, a good airfield performance and a typical range of 1,200nm (2,222km), although the aircraft was to be competitive on stage lengths as short as 350nm (648km). Cat III automatic landing capability was also to be built in. The podded underwing layout for the engines, although little differ-ent in performance from a rear-fuselage location, was consid-ered more suitable for the different power plants then under development. In terms of cockpit and cabin interior layouts, a considerable amount of input was provided at the detail design stage by the prospective customer airlines from the Atlas grouping, which provided some 100 specialists to work along-side the designers for one year.

The conventional layout of the Airbus, chosen to minimise development costs and risks, provided few obvious visual clues to the advanced aerodynamic concepts incorporated in its design. This particularly applied to the high-speed 28° swept wing, which had been under development at Hawker Siddeley for nearly nine years and was notable for the application of the rear-loading principle. This new technique, partially exploited on the Trident aircraft, was designed to considerably increase lift over the rear portion of the aerofoil before the onset of flow separation and buffet, through extending the higher-velocity

RIGHT: Airbus Industrie's Toulouse headquarters.

BELOW: An unusual hazard on the roads — a convoy of early Airbus sections trundling through the French countryside on their way to Toulouse.

ABOVE: In February 1972 the first fuselage section of the A300B prototype is transported inside the transport Guppy.

RIGHT: The French-built Airbus Skylink 3 was one of four Guppies transporting Airbus assemblies between the various production plants.

BELOW: A300B model in wind testing.

flow further aft on the upper surfaces. Significant advances were also made in the wing leading-edge design, which provided a more efficient compromise between high- and low-speed requirements. A desire for mechanical simplicity ruled out the more usual triple-slotted trailing-edge flap arrangement, in favour of tabbed Fowler flaps, which had been used satisfactorily by Bréguet on the Atlantic and by Sud-Aviation on the Caravelle. Hawker Siddeley's aerodynamic research work was backed up by over 3,500 hours of wind tunnel testing in research centres in Britain, France, Germany and the Netherlands, which identified the critical areas at an early stage in the Airbus design. That work led to a number of changes of which the most significant was the lengthening of the rear fuselage. This permitted the use of a smaller, and therefore lighter, fin and tailplane, resulting in lower trim and yaw drag, with a consequent improvement in overall performance.

Britain's wavering commitment to the project, compounded by Rolls-Royce forging ahead with the RB211 for the Lockheed TriStar, had already convinced Airbus Industrie to go with the new General Electric CF6 turbofan, which was being developed for the McDonnell Douglas DC-10 tri-jet. Plans were, therefore, being drawn up to build four development aircraft, designated A300B1, around this specification, although until Britain's withdrawal and even afterwards, the smaller Rolls-Royce RB211 engine had been an option for the fourth aircraft, primarily to win the order from BEA. It was also considered that a choice of engines would increase the aircraft's market potential by up to 40 percent.

But BEA had different ideas. It was drawn to the widebody BAC 3-11, which was also powered by the RB 211 and had succeeded the cancelled 2-11. A larger aircraft with a similar specification to the Airbus, the 3-11 represented direct competition to the European aircraft. It made little commecial sense to have two competing aircraft in Europe, but BAC appeared to be winning the argument in the corridors of power, until a change of government turned everything upside down. The new Conservative administration had little time for aviation, and on 3 November 1970 aviation minister Fred Corfield announced that the government would not fund the 3-11, nor Britain's re-entry into Airbus, which until then had once again been under consideration. In the end, BEA got neither aircraft.

Across the Channel, Air France, anticipating an upswing in business traffic within Europe, was pressing for 24 more passengers in the Airbus. This was achieved by inserting two extra fuselage frames ahead and three aft of the wings, totalling 8ft 8½in (2.65m) and providing space for three more rows of eight seats each. The CF6-50 turbofan engine had already grown more powerful to suit the heavier aircraft, with the end result that Airbus could offer the more economical A300B2, as it was called, within days of the Air France request. It was then decided that the last two development aircraft were to be built to B2 configuration, with the last aircraft virtually to production standard. The length was finalised at 175ft 11in (53.60m), which provided up to 281 passengers in an eight-abreast, two-aisle configuration, with seating for 345 nine-abreast passengers possible. A generous underfloor cargo hold could accommodate up to 20 standard LD3 containers, or alternatively, four pallets and eight LD3s.

Airbus Model Evolution

Date	Type	Seats	Range	MTOW	Power Plant
09/67	A300	287	1,200nm	120,000kg	211.4kN (47,500lb) R-R RB207
04/68	A300	300	1,200nm	140,000kg	240.3kN (54,000lb) R-R RB207
11/68	A300	306	1,200nm	150,000kg	255.9kN (57,500lb) R-R RB207
12/68	A300B	250	1,200nm	125,000kg	191.4kN (43,000lb) R-R RB211-20 or -40
07/69	A300B1	259	1,200nm	132,000kg	218.0kN (49,000lb) GE CF6-50A
11/71	A300B2	281	1,200nm	137,000kg	227.0kN (51,000lb) GE CF6-50C
11/71	A300B4	281	2,100nm	150,000kg	227.0kN (51,000lb) GE CF6-50C

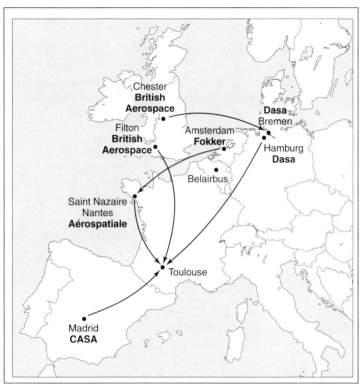

FIRST ORDERS

The long expected first order for the Airbus was finally signed on 9 November 1971, but the size of the order was disappointing, with Air France contracting to buy just six A300B2s and taking options on 10 more. Nevertheless, with Air France very much involved in the detail specification of the aircraft, the order provided a welcome boost for the project and the French flag-carrier showed its faith in the Airbus by building up its fleet on a regular basis over subsequent years. That same month, Airbus Industrie decided to offer an increased range model, the A300B4, to widen its market appeal and it was this, which became the standard model, although both were acceptable to the Atlas airlines, comprising Air France, Lufthansa, Alitalia, Sabena and allied Spanish flag-carrier Iberia. It was Iberia, which placed a contract for four A300B4s on 14 January 1972, becoming the second Airbus customer. Lufthansa, which had always pushed for a still

ABOVE: Airbus production flow.

BELOW: A300B prototype nearing completion.

smaller Airbus, continued its deliberations and did not announce its intention to buy until 19 December, with the contract for three A300B2s, plus four options signed on 7 May 1973.

The B4 differed only in having an additional centre wing tank of 28,000lb (12,500kg) capacity and a changed fuel-management system, which increased the take-off weight to 137 tonnes and the range to more than 2,000nm (3,700km). The physical dimensions were identical to the B2. An A300B3, basically a CF6-50C-powered variant with the higher B4 take-off

PROJECTED TWIN-ENGINED AIRBUS VERSIONS

		B1	B2	B3	B4
Wing span	(m)	44.83	44.83	44.83	44.83
	(ft/in)	147' 1"	147' 1"	147' 1"	147' 1"
Length	(m)	50.95	53.60	50.95	53.60
	(ft/in)	167' 2"	175' 10"	167' 2"	175' 10"
Accommodation	(typ)	257	281	257	281
Power Plant		CF6-50A	CF6-50C	CF6-50C	CF6-50C
MTOW	(kg)	132,000	137,000	148,500	150,000
	(lb)	291,000	302,030	327,400	330,690
MLW	(kg)	120,000	127,500	131,000	133,000
	(lb)	264,550	281,090	288,800	293,210
MZFW	(kg)	109,000	116,500	122,000	122,000
	(lb)	240,300	256,840	268,960	268,960
Typical range	(nm)	1,150	1,750	1,750	2,100
	(km)	2,130	3,240	3,240	3,885

		B5	B6	B7	B8
Wing span	(m)	44.83	44.83	44.83	45.75
	(ft)	147' 1"	147' 1"	147' 1"	150' 1"
Length	(m)	50.95	53.60	54.65	50.95
	(ft/in)	167' 2"	175' 10"	179' 4"	167' 2"
Accommodation	(typ)	cargo	cargo	290	257
Power Plant		CF6-50A	CF6-50C	RB	CF6-6
MTOW	(kg)	132,000	150,000	148,500	124,000
	(lb)	291,000	330,690	327,400	273,370
MLW	(kg)	120,000	133,000	131,000	114,000
	(lb)	264,550	293,210	288,800	351,325
MZFW	(kg)	109,000	116,500	120,000	105,000
	(lb)	240,300	256,840	264,550	231,480
Typical range	(nm)	1,150	2,100	1,600	1,000
	(km)	2,130	3,885	2,960	1,850

		B9*	B10**
Wing span	(m)	44.83	43.90
	(ft)	147 1"	144 0"
Length	(m)	57.27	46.66
	(ft/in)	187 11"	153 1"
Accommodation	(typ)	309	210
Power Plant		CF6-50C	CF6-50C
MTOW	(kg)	150,000	121,000
	(lb)	330,690	266,755
MLW	(kg)	135,000	115,000
	(lb)	297,620	253,530
MZFW	(kg)	125,000	95,000
	(lb)	275,575	209,450
Typical range	(nm)	2,160	1,750
	(km)	4,000	3,2,40

* later developed into the A330	MTOW	Maximum take-off weight
** became the A310	MLW	Maximum landing weight
	MZFW	Maximum zero-fuel weight

weight and an increased range of more than 3,240km (1,750nm), had been discarded early, and BEA was now also thought a prime candidate for the longer-range B4, rather than the specific Rolls-Royce-powered 290-seat A300B7, offered earlier.

WORK SHARE ALLOCATION AND CONSTRUCTION

The Airbus work was shared among the partners roughly in proportion to their financial holdings in the consortium, with Hawker Siddeley (later becoming part of British Aerospace) acting as a privileged subcontractor until Britain's re-joining in 1979. This approach clearly necessitated a wide geographical spread of production, although this was neither new, nor the most daunting aspect of construction, since still greater distances were overcome in the United States, even in non-co-operative ventures. But there was (and remains) a fundamental difference in the Airbus approach, which sets it apart from the rest. Right from its inception, Airbus had planned that all partners would contribute complete subassemblies in a 'ready-to-fly' condition, with all cables, pipe runs and equipment installed and checked, rather than merely becoming a manufacturer of parts and components. Not only was this a socio-economic master stroke, it also ensured that the huge reservoir of expertise — the Airbus partners were prime aircraft contractors

in their own right — was utilised to the full, and Airbus Industrie also benefited from pure research work sponsored by the partner countries. The result was that just 4 percent of man-hours required in building the Airbus models are spent on the final assembly line in Toulouse, which equates to around one-fifth of the work required by traditional methods.

There was one drawback to this philosophy, however. The large and heavy subassemblies basically ruled out transport by road on a long-term basis, while sections were also too large for moving by rail and sea transport considered too slow. But there, luck came its way and the unusual sight of large Airbus sections trundling through the French countryside on their way to Toulouse was quickly replaced by an even stranger apparition in the sky. In the mid-1960s, a timely confluence of an enterprising ex-USAF bomber pilot Jack Conroy, the availability of surplus Boeing Stratocruisers, and a NASA requirement for an outsize transport in support of the US space programme, gave birth to the Guppy series of outsize freighters. The lumbering giants were aptly named, since they looked as if they would be more at home in the deep sea.

But the cavernous fuselage and impressive 20-tonne payload

RIGHT: The first Airbus being rolled out of the assembly hall in Toulouse on 28 September 1972.

BELOW: Wing assembly at British Aerospace.

attracted the attention of Sud-Aviation during a visit of the ungainly aircraft to the Paris Air Show in 1969. A contract was placed with Aero Spacelines for two larger Allison-powered Super Guppy aircraft, based on the military KC-97 airframe. The first aircraft, designated Guppy 201, was delivered to Le Bourget on 29 September 1971, followed by the second a year later. The two aircraft were an immediate success and in subsequent years proved themselves as reliable workhorses.

As Airbus orders and production rates increased, however, it became clear that more were needed. Airbus Industrie acquired the production rights and commissioned UTA Industries at Le Bourget to build two more Guppy 201s, which entered service in 1982 and 1983. All four were flown by UTA subsidiary Aéromaritime until 1989, when Airbus Industrie took direct control of the operation. Known as Airbus Skylink 1, 2, 3 and 4, the four aircraft provided round-the-clock service, transporting wing, fuselage and tail assemblies from the various manufacturing sites of the partners in France, Germany, Spain and the UK. Each complete A300 required a total of eight flights, which amounted to some 45 hours in the air and a distance flown of nearly 7,000nm (13,000km). With its swing-out nose section opening through 110°, a 25ft 6in (7.77m) diameter cargo hold and a maximum payload of 22 tonnes, the Guppy 201 could easily accommodate the largest and heaviest A300 sections.

TAKING SHAPE

By Spring 1971, the prototype was beginning to take shape in the assembly hall at Toulouse. One complete airframe was also being built for structural testing for installation at the French government test centre Centre d'Essais Aéronautiques de Toulouse (CEAT), together with several fuselage sections for fatigue tests at Bremen, Hamburg and Ottobrünn near Munich from June 1972. In the meantime, the wings were being put together in Hawker Siddeley's Chester facility near Manchester in the UK, while work on the major fuselage sections was accelerated in France and Germany. The central wing box and nose section, manufactured by Aerospatiale at St Nazaire and Nantes-Bougenais, at first had to make the journey to Toulouse by road, before the Guppy took over later in 1971. The fuselage sections built by VFW-Fokker at Einswarden on the coast north of Bremen, had to be shipped by barge up the Weser river to the Lemwerder plant, from where they were flown to Toulouse, along with the rear fuselage sections built by MBB at Hamburg. The upper part of the centre fuselage, also built by MBB, first went to St Nazaire for integration with Aerospatiale's centre wing box

The Airbus wings, weighing a combined 20 tonnes, were taken from Chester to nearby Manchester airport by road on a specially-designed flat-bed trailer. Mounted horizontally to avoid any height restrictions en route and to ease loading onto

the Super Guppy, they were then flown to VFW-Fokker in Bremen, which fitted all moving surfaces manufactured elsewhere. Following Spain's decision to participate in the Airbus, the production of the tailplane and undercarriage doors was transferred from Germany to CASA's facility near Madrid, which later also started producing the forward passenger doors. The first fuselage assembly was completed on 19 November 1971, and the first pair of wings arrived at Toulouse four days later for mating with the fuselage. The CF6-50A engine followed on 2 April from Tula Vista, California, where it was podded up by Rohr Corporation, and fitted to the Airbus within two days. After an initial delivery programme from the United States, the General Electric CF6 engines were later assembled by Snecma near Paris, with some components built by Snecma and its German partner, MTU. The podding up was handled by Rohr in a specially-constructed building near the final assembly line.

Starting on 4 August 1972, the completed A300 undertook various ground runs, where it was taken to 100kts to test brakes and thrust reversers, before being introduced to the public, jointly with Concorde 02, during a rollout ceremony at Toulouse on 28 September. The enormous crowd, which included the French Prime Minister Pierre Messmer and the British Minister for Aerospace Michael Heseltine, had a grandstand view of two brand new aircraft designed and built in Europe, by Europeans. The significance of that occasion was not lost on those present. Pierre Messmer said that the Airbus took 'a great step forward by the creation of fully integrated international teams on all levels,' but also added that ambitions must now be directed towards a European dimension. The latter point was emphasised by Michael Heseltine who declared that 'collaboration on specific projects — important though it is — is not now enough. We must move from ad hoc collaboration on specific projects towards an integrated European aircraft industry; we face too much competition from the rest of the world to risk the prospect of competing with each other.'

UP, UP AND AWAY

Exactly one month later on a grey and windy morning of 28 October 1972, Airbus A300B1 F-WUAB was ready for its maiden flight. Piloted by Aerospatiale's senior test pilot Max Fischl, with Bernard Ziegler, then Head of Flight Test, as co-pilot, the Airbus lifted off the runway at Toulouse-Blagnac without a hitch and returned safely 1hour 23 minutes later. Also on board as flight engineers were Pierre Caneill and Günther Scherer, and Roméo Zinzoni as flight observer. Henri Ziegler, then President of Airbus Industrie, called the first flight 'a significant event in the history of the European Aeronautics industry. This flight,' he continued, 'is a major step on the way to a Europe that has come of age technologically, and beyond that, to a united Europe.'

During this first trouble-free flight, the Airbus climbed to 14,000ft (4,270m) and reached an indicated air speed (IAS) of 185kts (342km/h). Autopilot, undercarriage retraction and various flap settings were also tested. A second flight, of 2hr 30min duration, was made three days later, during which the Airbus extended the flight envelope to 20,000ft (6,100m) and 270kts. 11 flights, including one at night, had been completed by the end of November, totalling 32hr 30min, during which the entire flight envelope had been explored and handling, including single-engine performance, was reported as good.

Throughout the first half of 1973, the prototype, joined by the second A300B1 (F-WUAC) following its first 1hr 50min flight on 5 February, fitted in a busy demonstration programme, covering several European cities. This included a first flight outside France on 16 January, to Hawker Siddeley's Hatfield plant during the visit of a Chinese trade delegation. The first certification flight started in March, which was the 58th flight undertaken by the prototype, or as Airbus preferred to call it, the Airbus No 1. A slight hiccup occurred on 26 March 1973, when ice accumulation during icing trials at high altitude forced the shut-down of one of its CF6 engines and a three-day delay. However, the development programme

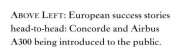

ABOVE LEFT: European success stories head-to-head: Concorde and Airbus A300 being introduced to the public.

LEFT: Airbus prototype on its first flight.

RIGHT: The flight crew on the first flight on the steps of the Airbus prototype. Captain in command Max Fischl is third from the right, with co-pilot Bernard Ziegler, the Head of Test Flight on the far right.

was then already one month ahead of schedule, helped by over 15,000 hours of wind tunnel testing and 300 hours of simulator work prior to flight testing. The prototype was also equipped with a Stanley Yankee rocket-assisted escape system for all five crew members, which speeded up the programme by permitting a more rapid approach to the potentially more dangerous areas of the flight envelope, while still maintaining adequate safety margins. The system was mounted in two groups behind the flightdeck opposite large jettisonable hatches. Continuous handrails were fitted to ease access during extreme conditions. They could be used at speeds up to 400kts (740km/h), but for lower speeds an escape route was provided through the forward cargo door.

On 8 May, the Airbus carried out its first automatic landing, using a system designed jointly by Sfena of France, Bodenseewerke of Germany, and the UK's Smiths Instruments. By the middle of the month, a diving speed of Mach 0.937 and corrected air speed of 432kts, signalled clearance of the last area of the flight envelope. With excellent stability and performance generally better than predicted, few changes were required to

LEFT: A striking view of the first Airbus on flight trials.

BELOW: The first two A300Bs on test flights over the Pyranees in 1973.

the aircraft during the test programme. A vortex generator was added in front of the ailerons to eliminate a slight lateral unsteadiness, and a small fence was added to the wing leading edge to achieve a more positive nose-down attitude in a clean stall at aft cg positions. Lift co-efficients were some 13 percent higher than predicted on take-off (9 percent on landing), which permitted a reduction of flap positions from five to three and a consequent simplification of standard single- and twin-engined approach procedures. The wing buffet boundary was 610m (2,000ft) better than predicted. A re-design of the intake and the use of fixed doors became necessary on the Garrett auxiliary power unit (APU), to ensure trouble-free relight clearance up to 35,000ft and at speeds above Mach 0.75.

DEMONSTRATING THE NEW AIRBUS TO THE WORLD

Airbus then took the courageous decision to take the prototype on a lengthy world tour to stimulate interest in its new airliner. While many considered this decision a foolhardy exercise so early in the aircraft's flight development programme, the lack of orders and a persistent scepticism among the world's airlines virtually forced Airbus into taking this unusual course of action. But with typical French panache, Airbus turned on the style, serving champagne to the thousands of visitors who ven-

tured aboard on the many stops along the way, with stewards and stewardesses dressed by Ted Lapidus and Courreges respectively. The prototype left Toulouse on 15 September 1973 and spent the next few days at the air show at Sao Paulo in Brazil, before flying on to Rio de Janeiro, Brasilia, Port of Spain, Miami, Mexico City, Chicago, Cincinnati, St Louis, Boston, New York, Kingston, Caracas, Washington and Montreal.

The prototype performed flawlessly, except for one incident, which had its beginning across the South Atlantic when one of the General Electric CF6 engines sustained some minor foreign object damage to a number of turbine blades, confirmed during a boroscope inspection at Mexico City. While the damage was considered insufficient to halt the tour, it was, nevertheless, decided to make an engine change at Chicago, the next port of call. After take-off with a heavy payload on a hot day at Mexico City, an oil leak was discovered, which was later traced to a sealing ring damaged during the reassembly of the engine. As a precautionary measure, the suspect engine was shut down and the aircraft landed safely back at the airport. At Chicago O'Hare a new engine, brought over from Toulouse aboard a

LEFT: Cockpit view of the third Airbus prototype.

BELOW LEFT: Iberia became the second customer for the A300. Its order was closely linked to Spain's participation in the Airbus programme. *R.L.Ward*

BELOW: Rio de Janeiro forms a splendid backdrop to the A300 prototype on its demonstration trip to the Americas in September 1973.

Transall, was fitted smoothly in a six-hour overnight operation at the United Air Lines workshops. These events demonstrated that fears about engine-out performance of the new twin-engine widebody aircraft were exaggerated, and that Airbus' product support was a match for the established manufacturers. Although none of the tours generated immediate firm orders, the flights had served to establish the aircraft's credibility, especially with a sceptical US airline fraternity.

On its return on 18 October via Gander, London and Paris, it continued its flight test programme, with now three aircraft — the third Airbus and first A300B2 (F-WUAD) had flown on 28 June — passing 1,000 development hours on over 400 flights exactly one year after the first flight. On 31 October, Airbus No 1 was off again, this time to India, where it made five demonstration flights at Bombay and Delhi, taking in visits to Athens, Tehran, Belgrade and Amsterdam. Syrian Arab Airlines received a demonstration on 15 November. The fourth and final aircraft involved in the development programme, an A300B2 registered F-WUAA, flew on 20 November 1973, freeing others for more sales tours. Airbus No.1 then logged 51 flight hours on a tour to southern Africa, where it was presented at Johannesburg and carried out hot-and-high trials at Windhoek and Kinshasa, before returning to Toulouse via Casablanca, Algiers, Tunis and Rome. Airbus No 2 meanwhile was engaged in cold weather trials in Helsinki and at Rovaniemi on the Arctic Circle.

Structural tests were completed on 22 October 1973, a month prior to the first flight of the fourth and last develop-

A300 FLIGHT TEST PROGRAMME ALLOCATION

A/C	Model	First Flight	Reg	Hours	Tasks
0001	B1	28/10/1972	F-WUAB	580	Flight characteristics, systems
0002	B1	05/02/1973	F-WUAC	500	Power plant, APU, de-icing
0003	B2	28/06/1973	F-WUAD	300	Autopilot and performance
0004	B2	20/11/1973	F-WUAA	210	Automatic landing
TOTAL				1,590	

ment aircraft, which was also the first production A300B2. At the beginning of January 1974, Airbus Industrie moved its headquarters from Paris to Toulouse, to be closer to the action. The final B2 certification flight was made on 31 January 1974, with simultaneous French and German certification to Cat II being obtained on 15 March 1974, two months ahead of the schedule drawn up five years before, on 28 May 1969. The four aircraft had spent a total of 1,585 hours in the air, which included 1,205 hours on development and certification work, and 380 hours on route proving, training and demonstrations to potential customers. Only four flights of those allocated, had to be cancelled, because of minor engine and brake problems. US certification by the Federal Aviation Administration (FAA) followed on 30 May, and on 30 September that same year, the Airbus was cleared for Cat IIIa autoland operations. The latter was achieved after 1,282 auto-approaches and 840 automatic landings.

BELOW: Air France's first A300B2-1A photographed in April 1974 ready and waiting proir to first delivery.

3 TECHNICAL SPECIFICATION

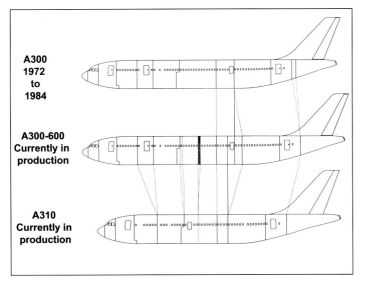

ABOVE: A300/A300-800 and A310 segment differences.

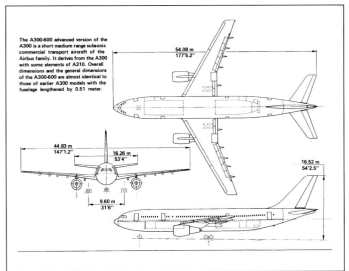

ABOVE RIGHT: General arrangement of the A300B4-600.

MODEL		B1	B2	B2K	B2-100	B2-200
External dimensions						
Wingspan	(m)	44.84	44.84	44.84	44.84	44.84
	(ft in)	147'1''	147'1''	147'1''	147'1''	147'1''
Length overall	(m)	50.97	53.62	53.62	53.62	53.62
	(ft in)	167'2''	175'1''	175'1''	175'1''	175'1''
Height overall	(m)	16.53	16.53	16.53	16.53	16.53
	(ft in)	54'3''	54'3''	54'3''	54'3''	54'3''
Tailplane span	(m)	16.94	16.94	16.94	16.94	16.94
	(ft in)	55'7''	55'7''	55'7''	55'7''	55'7''
Max fuselage dia	(m)	5.64	5.64	5.64	5.64	5.64
	(ft in)	18'6''	18'6''	18'6''	18'6''	18'6''
Wheel track	(m)	9.60	9.60	9.60	9.60	9.60
	(ft in)	31'6''	31'6''	31'6''	31'6''	31'6''
Wheelbase	(m)	17.57	18.60	18.60	18.60	18.60
	(ft in)	57'7''	61'0''	61'0''	61'0''	61'0''
Internal dimensions						
Main cabin length	(m)	36.50	39.15	39.15	39.15	39.15
	(ft in)	119'9''	128'6''	128'6''	128'6''	128'6''
Max cabin width	(m)	5.28	5.28	5.28	5.28	5.28
	(ft in)	17'4''	17'4''	17'4''	17'4''	17'4''
Max cabin height	(m)	2.54	2.54	2.54	2.54	2.54
	(ft in)	8'4''	8'4''	8'4''	8'4''	8'4''
Areas						
Wing, gross	(sq m)	260.0	260.0	260.0	260.0	260.0
	(sq ft)	2,799	2,799	2,799	2,799	2,799
Leading-edge	(sq m)	30.51	30.51	30.51	30.51	30.51
slats	(sq ft)	328.4	328.4	328.4	328.4	328.4
Kruger flaps	(sq m)	—	—	1.12	1.12	1.12
	(sq ft)	—	—	12.0	12.0	12.0

MODEL		B1	B2	B2K	B2-100	B2-200
Trailing-edge	(sq m)	46.60	46.60	46.60	46.60	46.60
flaps	(sq ft)	501.6	501.6	501.6	501.6	501.6
Ailerons	(sq m)	12.79	12.79	12.79	12.79	12.79
	(sq ft)	137.6	137.6	137.6	137.6	137.6
Spoilers	(sq m)	5.4	5.4	5.4	5.4	5.4
	(sq ft)	58.1	58.1	58.1	58.1	58.1
Airbrakes	(sq m)	8.10	8.10	8.10	8.10	8.10
	(sq ft)	87.2	87.2	87.2	87.2	87.2
Fin	(sq m)	45.24	45.24	45.24	45.24	45.24
	(sq ft)	486.9	486.9	486.9	486.9	486.9
Rudder	(sq m)	13.57	13.57	13.57	13.57	13.57
	(sq ft)	146.1	146.1	146.1	146.1	146.1
Tailplane	(sq m)	69.45	69.45	69.45	69.45	69.45
	(sq ft)	748.1	747.6	747.6	747.6	747.6
Elevators	(sq m)	17.85	17.85	17.85	17.85	17.85
	(sq ft)	192.1	192.1	192.1	192.1	192.1
Accommodation						
Passengers	(typ/max)	220/302	251/345	251/345	251/345	251/345
Cargo volume	(cu m)	110.0	140.0	140.0	140.0	140.0
	(cu ft)	3,885	4,945	4,945	4,945	4,945
Power plant	(no/type)	2 x CF6-50A	2 x CF6-50C	2 x CF6-50C	2 x CF6-50C	2 x CF6-50C
Thrust (each)	(kN)	218	222.5	227.0	227.0	227.0
	(lb)	49,000	50,000	51,000	51,000	51,000
Weights and loadings						
Max TO wt	(kg)	120,000	137,000	137,000	137,000	142,000
	(lb)	264,550	302,030	302,030	302,030	313,050

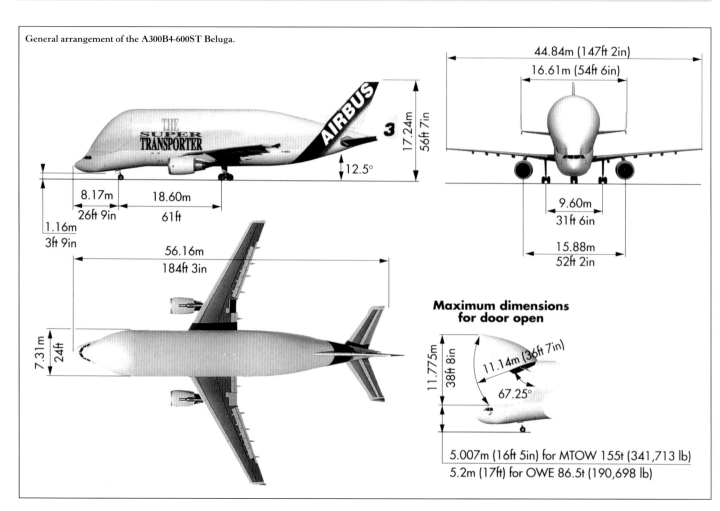

General arrangement of the A300B4-600ST Beluga.

MODEL		B1	B2	B2K	B2-100	B2-200
Max landing wt	(kg)	114,000	127,500	127,500	127,500	130,000
	(lb)	251,325	281,090	281,090	281,090	286,600
Max zero-fuel wt	(kg)	96,000	116,500	116,500	116,500	120,500
	(lb)	211,640	256,835	256,835	256,835	265,655
Operating wt empty	(kg)	69,000	85,910	85,910	89,100	89,100
	(lb)	152,120	189,400	189,400	196,430	196,430
Max payload	(kg)	28,970	34,590	34,590	34,600	34,600
	(lb)	63,870	76,260	76,260	76,280	76,280
Fuel capacity	(kg)	35,315	35,315	35,315	35,315	35,315
	(lb)	77,855	77,855	77,855	77,855	77,855
Performance						
Max permitted	(kts)	360	345	345	345	345
operating	(km/h)	666	638	638	638	638
speed (Vmo)	(mph)	414	396	396	396	396
Max op speed (Mmo)	(Mach)	0.84	0.86	0.86	0.86	0.86
Typical cruise speed	(kts)	457	457	457	457	457
	(km/h)	845	845	845	845	845
	(mph)	525	525	525	525	525
Approach speed	(kts)	128	131	131	131	131
	(km/h)	237	243	243	243	243
	(mph)	147	151	151	151	151
Max operating alt	(m)	10,675	10,675	10,675	10,675	10,675
	(ft)	35,000	35,000	35,000	35,000	35,000
TO field length	(m)	2,250	1,951	1,951	1,753	1,753
	(ft)	7,380	6,400	6,400	5,750	5,750
Landing field length	(m)	1,780	1,630	1,630	1,494	1,463
	(ft)	5,840	5,350	5,350	4,900	4,800
Range (typ payload)	(nm)	1,150	1,750	1,850	1,850	
	(km)	2,130	3,240	3,425	3,425	
	(miles)	1,320	2,010	2,125	2,125	

MODEL		B4-100	B4-120	B4-200	B4-220	B4-320
External dimensions						
Wingspan	(m)	44.84	44.84	44.84	44.84	44.84
	(ft in)	147'1"	147'1"	147'1"	147'1"	147'1"
Length overall	(m)	53.62	53.62	53.62	53.62	53.62
	(ft in)	175'1"	175'1"	175'1"	175'1"	175'1"
Height overall	(m)	16.53	16.53	16.53	16.53	16.53
	(ft in)	54'3"	54'3"	54'3"	54'3"	54'3"
Tailplane span	(m)	16.94	16.94	16.94	16.94	16.94
	(ft in)	55'7"	55'7"	55'7"	55'7"	55'7"
Max fuselage dia	(m)	5.64	5.64	5.64	5.64	5.64
	(ft in)	18'6"	18'6"	18'6"	18'6"	18'6"
Wheel track	(m)	9.60	9.60	9.60	9.60	9.60
	(ft in)	31'6"	31'6"	31'6"	31'6"	31'6"
Wheelbase	(m)	18.60	18.60	18.60	18.60	18.60
	(ft in)	61'0"	61'0"	61'0"	61'0"	61'0"
Internal dimensions						
Main cabin length	(m)	39.15	39.15	39.15	39.15	39.15
	(ft in)	128'6"	128'6"	128'6"	128'6"	128'6"
Max cabin width	(m)	5.28	5.28	5.28	5.28	5.28
	(ft in)	17'4"	17'4"	17'4"	17'4"	17'4"
Max cabin height	(m)	2.54	2.54	2.54	2.54	2.54
	(ft in)	8'4"	8'4"	8'4"	8'4"	8'4"

MODEL		B4-100	B4-120	B4-200	B4-220	B4-320
Areas						
Wing, gross	(sq m)	260.0	260.0	260.0	260.0	260.0
	(sq ft)	2,799	2,799	2,799	2,799	2,799
Leading-edge slats	(sq m)	30.51	30.51	30.51	30.51	30.51
	(sq ft)	328.4	328.4	328.4	328.4	328.4
Krüger flaps	(sq m)	1.12	1.12	1.12	1.12	1.12
	(sq ft)	1.12	1.12	12.0	12.0	12.0
Trailing-edge flaps	(sq m)	46.60	46.60	46.60	46.60	46.60
	(sq ft)	501.6	501.6	501.6	501.6	501.6
Ailerons	(sq m)	12.79	12.79	12.79	12.79	12.79
	(sq ft)	137.6	137.6	137.6	137.6	137.6
Spoilers	(sq m)	5.4	5.4	5.4	5.4	5.4
	(sq ft)	58.1	58.1	58.1	58.1	58.1
Airbrakes	(sq m)	8.10	8.10	8.10	8.10	8.10
	(sq ft)	87.2	87.2	87.2	87.2	87.2
Fin	(sq m)	45.24	45.24	45.24	45.24	45.24
	(sq ft)	486.9	486.9	486.9	486.9	486.9
Rudder	(sq m)	13.57	13.57	13.57	13.57	13.57
	(sq ft)	146.1	146.1	146.1	146.1	146.1
Tailplane	(sq m)	69.45	69.45	69.45	69.45	69.45
	(sq ft)	747.6	747.6	747.6	747.6	747.6
Elevators	(sq m)	17.85	17.85	17.85	17.85	17.85
	(sq ft)	192.1	192.1	192.1	192.1	192.1
Accommodation						
Passengers	(typ/max)	251/345	251/345	251/345	251/345	251/345
Cargo volume	(cu m)	140.0	140.0	140.0	140.0	140.0
	(cu ft)	4,945	4,945	4,945	4,945	4,945
Power plant	(no/type)	2xCF6-50C2	2xJT9D-59A	2xCF6-50C2	2xJT9D-59A	2xJT9D-59A
Thrust (each)	(kN)	233.6	235.9	233.6	235.9	235.9
	(lb)	52,500	53,000	52,500	53,000	53,000
Weights and loadings						
Max take-off wt	(kg)	157,500	157,500	165,000	165,000	165,000
	(lb)	347,225	347,225	363,760	363,760	363,760
Max landing wt	(kg)	134,000	136,000	134,000	136,000	136,000
	(lb)	295,420	299,825	295,420	299,825	299,825
Max zero-fuel wt	(kg)	124,000	126,000	124,000	126,000	126,000
	(lb)	273,370	277,780	273,370	277,780	277,780
Operating wt empty	(kg)	90,700	89,300	91,000	89,700	89,700
	(lb)	199,960	196,870	200,620	197,755	197,755
Max payload	(kg)	35,900	34,700	34,300	34,300	34,700
	(lb)	79,145	76,500	75,620	75,620	76,500
Fuel capacity	(kg)	49,760	49,760	49,760	49,760	49,760
	(lb)	109,700	109,700	109,700	109,700	109,700
Performance						
Max operating	(kts)	345	345	345	345	345
speed (Vmo)	(km/h)	638	638	638	638	638
	(mph)	396	396	396	396	396
Max op speed (Mmo)	(Mach)	0.82	0.82	0.82	0.82	0.82
Typical cruise speed	(kts)	457	457	457	457	457
	(km/h)	845	845	845	845	845
	(mph)	525	525	525	525	525
Approach speed	(kts)	132	132	132	132	132
	(km/h)	244	244	244	244	244
	(mph)	152	152	152	152	152

ABOVE: **A300 fuselage sections awaiting final assembly.** *Günter Endres*

Max operating altitude	(m)	10,675	10,675	10,675	10,675	10,675
	(ft)	35,000	35,000	35,000	35,000	35,000
TO field length	(m)	2,332	2,286	2,972	2,774	2,774
	(ft)	7,650	7,500	9,750	9,100	9,100
Landing field length	(m)	1,660	1,660			
	(ft)	5,445	5,445			
Range (typ payload)	(nm)	3,400	3,500	3,400	3,500	3,400
	(km)	6,290	6,475	6,290	6,475	6,290
	(miles)	3,910	4,020	3,910	4,020	3,910
MODEL		**C4-200**	**F4-200**	**B4-600**	**B4-620**	**C4-620***
External dimensions						
Wingspan	(m)	44.84	44.84	44.84	44.84	44.84
	(ft in)	147'1"	147'1"	147'1"	147'1"	147'1"
Length overall	(m)	53.62	53.62	54.08	54.08	54.08
	(ft in)	175'1"	175'1"	177'5"	177'5"	177'5"
Height overall	(m)	16.53	16.53	16.53	16.53	16.53
	(ft in)	54'3"	54'3"	54'3"	54'3"	54'3"
Tailplane span	(m)	16.94	16.94	16.26	16.26	16.26
	(ft in)	55'7"	55'7"	53'4"	53'4"	53'4"
Max fuselage dia	(m)	5.64	5.64	5.64	5.64	5.64
	(ft in)	18'6"	18'6"	18'6"	18'6"	18'6"
Wheel track	(m)	9.60	9.60	9.60	9.60	9.60
	(ft in)	31'6"	31'6"	31'6"	31'6"	31'6"
Wheelbase	(m)	18.60	18.60	18.60	18.60	18.60
	(ft in)	61'0"	61'0"	61'0"	61'0"	61'0"
Internal dimensions						
Main cabin length	(m)	39.15	39.15	40.21	40.21	40.21
	(ft in)	128'6"	128'6"	131'11"	131'11"	131'11"
Max cabin width	(m)	5.28	5.28	5.28	5.28	5.28
	(ft in)	17'4"	17'4"	17'4"	17'4"	17'4"
Max cabin height	(m)	2.54	2.54	2.54	2.54	2.54
	(ft in)	8'4"	8'4"	8'4"	8'4"	8'4"

Model		C4-200	F4-200	B4-600	B4-620	C4-620*
Areas						
Wing, gross	(sq m)	260.0	260.0	260.0	260.0	260.0
	(sq ft)	2,799	2,799	2,799	2,799	2,799
Leading-edge slats	(sq m)	30.51	30.51	30.30	30.30	30.30
	(sq ft)	328.4	328.4	326.1	326.1	326.1
Krüger flaps	(sq m)	1.12	1.12	1.12	1.12	1.12
	(sq ft)	12.0	12.0	12.0	12.0	12.0
Trailing-edge flaps	(sq m)	46.60	46.60	47.30	47.30	47.30
	(sq ft)	501.6	501.6	509.1	509.1	509.1
Ailerons	(sq m)	12.79	12.79	7.06	7.06	7.06
	(sq ft)	137.6	137.6	76.0	76.0	76.0
Spoilers	(sq m)	5.40	5.40	5.40	5.40	5.40
	(sq ft)	58.1	58.1	58.1	58.1	58.1
Airbrakes	(sq m)	8.10	8.10	12.59	12.59	12.59
	(sq ft)	87.2	87.2	135.2	135.2	135.2
Fin	(sq m)	45.24	45.24	45.20	45.20	45.20
	(sq ft)	486.9	486.9	486.5	486.5	486.5
Rudder	(sq m)	13.57	13.57	13.57	13.57	13.57
	(sq ft)	146.1	146.1	146.1	146.1	146.1
Tailplane	(sq m)	69.45	69.45	44.80	44.80	44.80
	(sq ft)	747.6	747.6	482.2	482.2	482.2
Elevators	(sq m)	17.85	17.85	19.20	19.20	19.20
	(sq ft)	192.1	192.1	206.7	206.7	206.7
Accommodation						
Passengers	(typ/max)	145/315		266/361	266/361	266/361
Cargo volume	(cu m)	140.0	319.0	147.4	147.4	147.4
	(cu ft)	4,945	11,267	5,206	5,206	5,206
Power plant	(no/type)	2xCF6-50C2	2xCF6-50C2	2xCF6-80C2A1	JT9D-7R4H1	JT9D-7R4H1
Thrust (each)	(kN)	233.6	233.6	262.5	249.2	249.2
	(lb)	52,500	52,500	59,000	56,000	56,000
Weights and loadings						
Max TO wt	(kg)	165,000	165,000	165,000	165,000	170,500
	(lb)	363,765	363,765	363,765	363,765	375,900
Max landing wt	(kg)	136,000	136,000	138,000	138,000	140,000
	(lb)	299,830	299,830	304,240	304,240	308,650
Max zero-fuel wt	(kg)	126,000	126,000	130,000	130,000	130,000
	(lb)	277,780	277,780	286,600	286,600	286,600

BELOW: Final assembly is taking only 4 percent of total man hours. *Günter Endres*

MODEL		C4-200*	F4-200	B4-600	B4-620	C4-620*
Operating wt empty	(kg)	89,200	79,400	90,115	90,065	93,475
	(lb)	196,650	175,045	198,665	198,565	206,075
Max payload	(kg)	35,000	45,000	39,885	39,995	36,525
	(lb)	77,160	99,210	87,930	88,175	80,525
Fuel capacity	(kg)	49,760	49,760	49,760	49,760	49,760
	(lb)	109,700	109,700	109,700	109,700	109,700
Performance						
Max operating	(kts)	345	345	335	335	335
speed (Vmo)	(km/h)	638	638	620	620	620
	(mph)	396	396	385	385	385
Max op speed (Mmo)	(Mach)	0.82	0.82	0.82	0.82	0.82
Typical cruise speed	(kts)	457	457	472	472	472
	(km/h)	845	845	873	873	873
	(mph)	525	525	543	543	543
Approach speed	(kts)	132	132	135	135	135
	(km/h)	244	244	250	250	250
	(mph)	152	152	155	155	155
Max op altitude	(m)	10,675	10,675	12,200	12,200	12,200
	(ft)	35,000	35,000	40,000	40,000	40,000
TO field length	(m)	2,750	2,750	2,378	2,270	2,270
	(ft)	9,020	9,020	7,800	7,450	7,450
Landing field length	(m)	1,660	1,660	1,536	1,536	1,536
	(ft)	5,445	5,445	5,040	5,040	5,040
Range (typ payload)	(nm)	2,500	2,500	3,650	3,700	2,650
	(km)	4,625	4,625	6,755	6,845	4,905
	(miles)	2,875	2,875	4,195	4,250	3,045

MODEL		F4-600	B4-600R	B4-620R	F4-600R	B4-600ST
External dimensions						
Wingspan	(m)	44.84	44.84	44.84	44.84	44.84
	(ft in)	147'1"	147'1"	147'1"	147'1"	147'1"
Length overall	(m)	54.08	54.08	54.08	54.08	56.16
	(ft in)	177'5"	177'5"	177'5"	177'5"	184'3"
Height overall	(m)	16.53	16.53	16.53	16.53	17.23
	(ft in)	54'3"	54'3"	54'3"	54'3"	56'6"
Tailplane span	(m)	16.26	16.26	16.26	16.26	16.61
	(ft in)	53'4"	53'4"	53'4"	53'4"	54'6"
Max fuselage dia	(m)	5.64	5.64	5.64	5.64	5.64
	(ft in)	18'6"	18'6"	18'6"	18'6"	18'6"
Wheel track	(m)	9.60	9.60	9.60	9.60	9.60
	(ft in)	31'6"	31'6"	31'6"	31'6"	31'6"
Wheelbase	(m)	18.60	18.60	18.60	18.60	18.60
	(ft in)	61'0"	61'0"	61'0"	61'0"	61'0"
Internal dimensions						
Main cabin length	(m)	40.21	40.21	40.21	40.21	37.70
	(ft in)	131'11"	131'11"	131'11"	131'11"	123'8"
Max cabin width	(m)	5.28	5.28	5.28	5.28	7.10
	(ft in)	17'4"	17'4"	17'4"	17'4"	23'3"
Max cabin height	(m)	2.54	2.54	2.54	2.54	7.10
	(ft in)	8'4"	8'4"	8'4"	8'4"	23'3"
Areas						
Wing, gross	(sq m)	260.0	260.0	260.0	260.0	260.0
	(sq ft)	2,799	2,799	2,799	2,799	2,799

Model		F4-600	B4-600R	B4-620R	F4-600R	B4-600ST
Leading-edge slats	(sq m)	30.30	30.30	30.30	30.30	30.30
	(sq ft)	326.1	326.1	326.1	326.1	326.1
Krüger flaps	(sq m)	1.12	1.12	1.12	1.12	1.12
	(sq.ft)	12.0	12.0	12.0	12.0	12.0
Trailing-edge flaps	(sq m)	47.30	47.30	47.30	47.30	47.30
	(sq ft)	509.1	509.1	509.1	509.1	509.1
Ailerons	(sq m)	7.06	7.06	7.06	7.06	7.06
	(sq ft)	76.0	76.0	76.0	76.0	76.0
Spoilers	(sq m)	5.40	5.40	5.40	5.40	5.40
	(sq ft)	58.1	58.1	58.1	58.1	58.1
Airbrakes	(sq m)	12.59	12.59	12.59	12.59	12.59
	(sq ft)	135.5	135.5	135.5	135.5	135.5
Fin	(sq m)	45.20	45.20	45.20	45.20	
	(sq ft)	486.5	486.5	486.5	486.5	
Rudder	(sq m)	13.57	13.57	13.57	13.57	
	(sq ft)	146.1	146.1	146.1	146.1	
Tailplane	(sq m)	44.80	44.80	44.80	44.80	
	(sq ft)	482.2	482.2	482.2	482.2	
Elevators	(sq m)	17.85	17.85	17.85	17.85	
	(sq ft)	192.1	192.1	192.1	192.1	
Accommodation						
Passengers	(typ/max)		266/361	266/361		
Cargo volume	(cu m)	350.4	147.4	147.4		1,400
	(cu ft)	12,376	5,206	5,206		49,448
Power plant (CF6s)	(no/type)	2x-80C2A5	2x-80C2A5	2xPW4158	2x-80C2A5	2x-80C2A8
Thrust (each)	(kN)	273.7	273.7	258.1	273.7	262.4
	(lb)	61,500	61,500	58,000	61,500	59,000
Weights and loadings						
Max TO wt	(kg)	165,000	170,500	170,500	170,500	153,000
	(lb)	363,760	375,885	375,885	375,885	337,310
Max landing wt	(kg)	138,000	140,000	140,000	140,000	
	(lb)	308,645	308,645	308,645	308,645	
Max zero-fuel wt	(kg)	130,000	130,000	130,000	130,000	132,000
	(lb)	286,600	286,600	286,600	286,600	291,010
Operating wt empty	(kg)	80,600	89,800	90,965	81,300	86,500
	(lb)	177,690	197,975	200,550	179,235	190,700
Max payload	(kg)	49,400	38,960	39,040	48,700	45,500
	(lb)	108,910	85,890	86,070	107,365	100,310
Fuel capacity	(kg)	54,695	54,695	54,695	54,695	
	(lb)	120,580	120,580	120,580	120,580	
Performance						
Max operating	(kts)	335	335	335	335	
speed (Vmo)	(km/h)	620	620	620	620	
	(mph)	385	385	385	385	
Max op speed (Mmo)	(Mach)	0.82	0.82	0.82	0.82	0.70
Typical cruise speed	(kts)	472	472	472	472	
	(km/h)	873	873	873	873	
	(mph)	543	543	543	543	
Approach speed	(kts)	135	136	136	135	
	(km/h)	250	252	252	252	
	(mph)	155	157	157	157	
Max op altitude	(m)	12,200	12,200	12,200	12,2000	
	(ft)	40,000	40,000	40,000	40,000	
TO field length	(m)	2,408	2,408	2,362		
	(ft)	7,900	7,900	7,750		

MODEL		F4-600	B4-600R	B4-620R	F4-600R	B4-600ST
Landing field length	(m)	1,536	1,555	1,555		
	(ft)	5,040	5,100	5,100		
Range (typ payload)	(nm)	2,650	4,050	4,050	3,150	900
	(km)	4,905	7,495	7,495	5,830	1,665
	(miles)	3,045	4,655	4,655	3,620	1,035

MODEL		B2-100F**	B2-200F**	B4-100F**	B4-200F**
External dimensions					
Wingspan	(m)	44.84	44.84	44.84	44.84
	(ft in)	147'1"	147'1"	147'1"	147'1"
Length overall	(m)	53.62	53.62	53.62	53.62
	(ft in)	175'1"	175'1"	175'1"	175'1"
Height overall	(m)	16.53	16.53	16.53	16.53
	(ft in)	54'3"	54'3"	54'3"	54'3"
Tailplane span	(m)	16.94	16.94	16.94	16.94
	(ft in)	55'7"	55'7"	55'7"	55'7"
Max fuselage dia	(m)	5.64	5.64	5.64	5.64
	(ft in)	18'6"	18'6"	18'6"	18'6"
Wheel track	(m)	9.60	9.60	9.60	9.60
	(ft in)	31'6"	31'6"	31'6"	31'6"
Wheelbase	(m)	18.60	18.60	18.60	18.60
	(ft in)	61'0"	61'0"	61'0"	61'0"

BELOW: A300-600R nearing completion in final assembly. *Günter Endres*

MODEL		B2-100F**	B2-200F**	B4-100F**	B4-220F**
Internal dimensions					
Main cabin length	(m)	39.15	39.15	39.15	39.15
	(ft in)	128'6"	128'6"	128'6"	128'6"
Max cabin width	(m)	5.28	5.28	5.28	5.28
	(ft in)	17'4"	17'4"	17'4"	17'4"
Max cabin height	(m)	2.54	2.54	2.54	2.54
	(ft in)	8'4"	8'4"	8'4"	8'4"
Areas					
Wing, gross	(sq m)	260.0	260.0	260.0	260.0
	(sq ft)	2,799	2,799	2,799	2,799
Leading-edge slats	(sq m)	30.51	30.51	30.51	30.51
	(sq ft)	328.4	328.4	328.4	328.4
Kr‚ger flaps	(sq m)	1.12	1.12	1.12	1.12
	(sq ft)	12.0	12.0	12.0	12.0
Trailing-edge flaps	(sq m)	46.60	46.60	46.60	46.60
	(sq ft)	501.6	501.6	501.6	501.6
Ailerons	(sq m)	12.79	12.79	12.79	12.79
	(sq ft)	137.6	137.6	137.6	137.6
Spoilers	(sq m)	5.40	5.40	5.40	5.40
	(sq ft)	58.1	58.1	58.1	58.1
Airbrakes	(sq m)	8.10	8.10	8.10	8.10
	(sq ft)	87.2	87.2	87.2	87.2
Fin	(sq m)	45.24	45.24	45.24	45.24
	(sq ft)	486.9	486.9	486.9	486.9
Rudder	(sq m)	13.57	13.57	13.57	13.57
	(sq ft)	146.1	146.1	146.1	146.1
Tailplane	(sq m)	69.45	69.45	69.45	69.45
	(sq ft)	747.6	747.6	747.6	747.6
Elevators	(sq m)	17.85	17.85	17.85	17.85
	(sq ft)	192.1	192.1	192.1	192.1
Accommodation					
Passengers	(typ/max)	-	-	-	-
Cargo volume	(cu m)				
	(cu ft)				
Power plant	(no/type)	2xCF6-50C	2xCF6-50C	2xCF6-50C2	2xJT9D-59A
Thrust (each)	(kN)	227.0	227.0	233.6	235.9
	(lb)	51,000	51,000	52,500	53,000
Weights and loadings					
Max TO wt	(kg)	137,000	142,000	157,500	165,000
	(lb)	302,030	313,055	347,230	363,760
Max landing wt	(kg)	127,500	130,000	134,000	134,000
	(lb)	281,090	286,600	295,420	295,420
Max zero-fuel wt	(kg)	116,500	120,500	124,000	126,000
	(lb)	256,835	265,655	273,375	277,780
Operating wt empty	(kg)	79,600	79,600	81,500	81,760
	(lb)	175,485	175,485	179,675	180,250
Max payload	(kg)	40,900	40,900	42,600	44,200
	(lb)	90,170	90,170	93,915	97,445
Fuel capacity	(kg)	35,315	35,315	49,760	49,760
	(lb)	77,855	77,855	109,700	109,700
Performance					
Max operating	(kts)	345	345	345	345
speed (Vmo)	(km/h)	638	638	638	638

		B2-100F**	B2-200F**	B4-100F**	B4-220F**
	(mph)	396	396	396	396
MODEL		**B2-100F****	**B2-200F****	**B4-100F****	**B4-220F****
Max op speed (Mmo)	(Mach)	0.86	0.86	0.82	0.82
Typical cruise speed	(kts)	457	457	457	457
	(km/h)	845	845	845	845
	(mph)	525	525	525	525
Approach speed	(kts)	131	131	132	132
	(km/h)	242	242	244	244
	(mph)	151	151	152	152
Max op altitude	(m)	10,675	10,675	10,675	10,675
	(ft)	35,000	35,000	35,000	35,000
TO field length	(m)				
	(ft)				
Landing field length	(m)				
	(ft)				
Range (typ payload)	(nm)	1,350	1,350	2,200	2,650
	(km)	2,500	2,500	4,070	4,905
	(miles)	1,555	1,555	2,530	3,045

* passenger mode   ** all-freighter conversions

GENERAL ELECTRIC ENGINE SPECIFICATIONS

		CF6-50A	-50C	-50C1	-50C2	-80C2A1	-80C2A3	-80C2A5	-80C2A8
Take-off thrust	(kN)	218.0	227.0	233.6	233.6	262.5	267.9	273.7	262.5
	(lb)	49,000	51,000	52,500	52,500	59,000	60,200	61,500	59,000
Max cruise thrust*	(kN)	48.1	48.1	50.3	50.3	50.4	50.4	50.4	50.4
	(lb)	10,800	10,800	11,300	11,300	11,330	11,330	11,330	11,330
Bypass ratio		4.4	4.4	4.4	4.4	5.2	5.2	5.2	5.2
Fan diameter	(mm)	2,195	2,195	2,195	2,195	2.362	2.362	2.362	2.362
	(in)	86.4	86.4	86.4	86.4	93.0	93.0	93.0	93.0
Length	(mm)	4,648	4,648	4,648	4,648	4,267	4,267	4,267	4,267
	(in)	183	183	183	183	168	168	168	168
Dry weight	(kg)	3,956	3,956	3,956	3,956	4,144	4,144	4,144	4,144
	(lb)	8,721	8,721	8,721	8,721	9,135	9,135	9,135	9,135
SFC (T-O)	(mg/Ns)	10.90	11.05	11.05	11.05	9.46	9.32	9.63	9.74
	(lb/h/lb st)	0.385	0.390	0.390	0.390	0.334	0.329	0.340	0.344

* at 10,670m (35,000 feet), Mach 0.85

PRATT & WHITNEY ENGINE SPECIFICATIONS

		JT9D-59A	JT9D-7R4H1	PW4156	PW4158
Take-off thrust	(kN)	235.9	249.2	249.2	258.1
	(lb)	53,000	56,000	56,000	56,000
Max Cruise thrust*	(kN)	53.2	54.5		
	(lb)	11,950	12.250		
Bypass ratio		4.9	4.8	4.9	4.8
Fan diameter	(mm)	2,373	2,373	2,377	2,377
	(in)	93.4	93.4	93.6	93.6
Length	(mm)	3,358	3,371	3,901	3,901
	(in)	132.2	132.7	153.6	153.6
Dry weight	(kg)	4,146	4,023	4,273	4,273
	(lb)	9,140	8,870	9,420	9,420
SFC cruise*	(mg/Ns)	17.87	17.79	15.21	15.21
	(lb/h/lb st)	0.631	0.628	0.537	0.537

* at 10,670m (35,000 ft), Mach 0.85

TECHNICAL DESCRIPTION

The short/medium-range A300 was the first wide-body twin-engined aircraft to enter airline service. It evolved through the design stage to meet different and changing airline requirements, and later benefited through the application of new technologies as they became available. Only the first two aircraft were built in the original A300B1 configuration — one entering airline service — before a demand for more seats produced the A300B2, which became the first model to go into service with Air France in May 1974. Additional range was provided in the A300B4, first utilised by Germanair in May 1975. South African Airways became the first operator in November 1976 of the A300B2K, which features the Krüger flaps and increased take-off weights of the A300B4. This model was designated in 1978 as the A300B2-200, with the original A300B2 becoming the A300B2-100. At the same time, the B4 models were given the designations A300B4-100 and A300B4-200, the latter denoting aircraft with higher take-off weights. All models were powered by the General Electric CF6-50 turbofan engine, but the A300B2-300 introduced the Pratt & Whitney JT9D power plant. Hapag Lloyd took delivery of the first convertible A300C4-200 in January 1980, while Korean Air Lines followed in August 1986 with the A300F4-200 all-cargo variant. A few B2 and B4 aircraft were produced with a two-crew cockpit and were referred to as the A300FFCC.

The higher-capacity A300-600, which entered service with Saudia in June 1984, introduced advanced technologies, including a two-crew cockpit with CRT displays, and was followed into service by the extended-range A300-600R in April 1988. Both new versions were built with a choice of CF6-80C2 or JT9D/PW4000 engines. In April 1994, Federal Express accepted the first of a large fleet of the A300-600F freighters. Bearing little resemblance to the A300-600R on which it is based, is the A300-600ST Beluga Super Transporter, used to ferry Airbus parts to the assembly lines. The A300-600ST operated its first flight in that role in January 1996.

STRUCTURE

The A300 fuselage is of conventional aluminium alloy, semi-monocoque, fail-safe construction of circular cross-section. It is manufactured in nine structural sections (for ease of transportation) largely from sheet metal fabricated frames, open section stringers and skins. In general, the skins are formed from simple sheet metal, except in highly loaded areas, such as the centre fuselage at the wings and landing gear and the nose gear bay, where the skins are machined. Stringer attachment is by hot bonding or riveting. A considerable amount of composites has been introduced on the later A300-600/600R models, including floor struts and panels, spoilers, main landing gear and the complete fin box, adding up to 14,600lb (6.6 tonnes) and producing a weight saving of 1.5 tonnes. All areas of the fuselage are pressurised, except for the radome, the rear fuselage section (tailcone), the nose landing gear bay and the lower segment of the centre section, which includes the air conditioning, hydraulic and main landing gear bays. Each engine is supported on a pylon, which forms a fail-safe box-type frame, constructed of high tensile steel. The variable incidence tailplane is actuated by a fail-safe ball screwjack, which can be electrically or mechanically controlled. The structure is composed of a main fin box, a removable leading edge, rear shroud panels, two elevators and a tip fairing.

The Airbus wing has a thickness/chord ratio of 10.5 percent and a 28° sweepback and is made up of three main components providing a continuous and fail-safe two-spar box structure with machined skins and open-section stringers. It is built mostly of high-strength aluminium alloy, except for the spoilers, flap track fairings and wing fuselage fairings, which are of composite materials. The centre wing box is built integrally into the fuselage, to which the port and starboard cantilevered outer wing sections are attached. Movable surfaces on the leading edge include three-section slats with a cambered fence on each outboard segment and no cut out over the engine pylon, a Krüger flap at the wing root (except for the early A300B2), and a closing plate fitted in the area of the inboard slat. Trailing edge devices comprise three cambered tabless Fowler flaps, redesigned for the A300-600/600R; an all-speed aileron mounted behind the engine and doubling up as a flap; and two spoilers and five airbrakes on the upper trailing edge surfaces forward of the flaps, all of which can be used as lift dumpers. The flaps extend over 84 percent of each half span. The A300-600/600R wing was extensively modified to provide improved aerodynamics. The main elements include a new inner wing

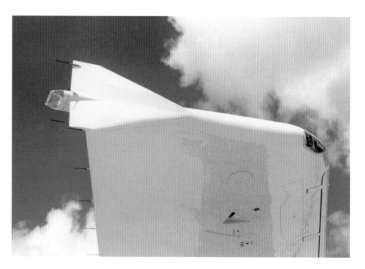

section, increased trailing-edge camber giving higher wing loading, deletion of the slat fence and outboard aileron, and an added wingtip fence for reduced cruise drag.

POWERPLANT

The A300 is in service with high bypass turbofan engines from the two major US manufacturers, General Electric and Pratt & Whitney, fitted on underwing pylons. While the first two A300B1 and two A300B2 used for development flying were equipped with the 49,000lb (218kN) thrust General Electric CF6-50A, the more powerful CF6-50C was fitted to subsequent B2-200 and B4-100 and B4-200 production models. Improved CF6-50C1, -C2 and C2R models were installed as they became available and to meet customer requirements. The A300B4-300 was the first to be powered by the Pratt & Whitney JT9D-59A, while the enhanced JT9D-7R4H1 later

BELOW FAR LEFT: Detail of the A300B4-600 weight reduction programme.

BELOW LEFT: Wingtip fence introduced on the A300-600/600R to reduce cruise drag *Günter Endres*

BELOW BOTTOM RIGHT: Close-up of the Pratt & Whitney PW4158 turbofan on wing.

BELOW AND RIGHT: Close-ups of the General Electric CF6-80C2. *Günter Endres*

became available on the new-technology A300-600, and the third-generation PW4000 on the extended-range A300-600R. Both -600 and -600R models have also been fitted with the General Electric CF6-80C (in -80C2A1, -80C2A3 and -80C2A5 versions), a re-design of the earlier CF6-50C. The A300-600ST Beluga is powered by the CF6-80C2A8.

The basic configuration of the 51,000lb (227.0kN) CF6-50C comprises a single-stage fan with integrally-mounted three-stage LP compressor acting as booster to the core engine. Both are driven by a four-stage LP turbine. The core engine consists of a 16-stage HP compressor, annular combustor, which can be removed with fuel nozzles in place, and a two-stage air-cooled HP turbine. Higher thrust became available with the CF6-50C1, rated at 52,500lb (233.5kN), while the CF6-50C2 features improved specific fuel consumption (sfc) and EGT margins, a new fan case and blades with better bird strike resistence. The CF6-80C2 is a complete re-design, providing a higher thrust range from 59,000lb (262.5kN) to 61,500lb (273.7kN), and lower specific fuel consumption and improved EGT margins. It has a new four-stage LP compressor, re-designed five-stage LP turbine, and 14-stage HP compressor with inlet guide vanes. The larger, single-stage fan features a Kevlar blade containment shroud.

The 53,000lb (235.85kN) thrust Pratt & Whitney JT9D-59A has a single-stage fan with integrally-mounted three-stage

LP compressor, four-stage LP turbine, 11-stage HP compressor, annular combustor, and two-stage HP turbine. The larger single-stage fan of the improved and higher 56,000lb (249.2kN) thrust JT9D-7R4H1 differs primarily in having 40 wide-chord blades (compared to 46 in the -59A), a four-stage LP compressor, improved combustor, single-crystal HP turbine blades, increased diameter LP turbine, and electronic supervisory fuel control. The third-generation PW4000 series, as well as providing higher thrust (58,000lb/258.1kN in the PW4158), is distinguished by a seven per cent reduction in sfc and improved maintainability through simplified construction. Other features are single-crystal turbine blades, aerodynamically-enhanced aerofoils, a more efficient Thermatic rotor, and full authority digital engine control (FADEC).

Early versions of both manufacturers' engines feature accessory drive gear boxes mounted externally on the fan case for ready access and a cool environment, but the later CF6-80C2, JT9D-7R4H1 and PW4000 series introduced a core-mounted gear box for cleaner aerodynamic lines and a more compact nacelle geometry. The three new-technology engines are also notable for a considerable improvement in noise attenuation through the use of composites in the nacelle and engine, with the associated weight savings. Smoke and emission levels have also been lowered. Air for starting the engine is supplied through a pneumatic manifold from an external high- pressure source, the onboard APU, or by cross-feed from the opposite engine.

Reverse thrust is provided by a cascade fan reverser system for each engine, actuated by pneumatic drive motors powered by engine compressor bleed air or pneumatic system air. The system of one engine is completely independent of the opposing engine. The fan reverser consists of translating sleeve blocker doors and fixed cascades, the latter tailored to generate effective retarding forces and to minimise exhaust gas re-injection at lower speeds. In the stowed position, the system forms a passage for the fan stream flow to the exhaust fan; while deployed, it provides reverse thrust, with thrust modulation being accomplished by power setting adjustments. Maximum thrust is permitted down to 60kts IAS. After inadvertent deployment in flight up to 300kts IAS, it is possible to re-stow the thrust reverser below 250kts IAS at 24,500ft (7,500m) altitude.

FUEL SYSTEM

Fuel capacity has been progressively increased through the A300 family range. The initial A300B2 was designed with two wing tanks, each divided into two compartments, with the inner holding about 78 percent of the total permissible capacity of 44,000 litres (11,610 US gallons; 9,679 Imp gallons). In the A300B4 version, the centre wing box provides additional tankage, which brought total allowable capacity to 62,000 litres (16,359 US gallons; 13,656 Imp gallons), also applicable to the A300-600/600R. As before, the two wing tanks are divided into an inner and outer compartment, each with a capacity of 17,570 litres (4,636 US gallons; 3,870 Imp gallons) and 4,630 litres (1,222 US gallons; 1,020 Imp gallons) respectively. The centre fuselage tank holds 17,600 litres (4,644 US gallons; 3,877 Imp gallons).

A feature of the most recent A300-600R is the installation of a 6,150 litres (1,620 US gallons; 1,355 Imp gallons) trim tank in the horizontal stabiliser. In addition to increasing the aircraft's range, a computerised fuel transfer system also provides active centre of gravity (c.g.) control for more efficient production of lift. The net result is a one per cent reduction in cruise drag and lower fuel burn. An optional extra fuel tank in the aft rear cargo hold can increase fuel capacity to 75,350 litres (19,992 US gallons; 16,597 Imp gallons) in the A300-600R.

The fuel system is designed for single-point refuelling/defuelling control by one operator from a panel located under the fuselage centre section. In the unlikely event of a complete AC power failure, suction feed is ensured up to an altitude of 20,000ft (6,100m). Two fuelling adaptors are provided on each wing, situated on the underside forward of the front spar outboard of the engine. Normally, each engine is supplied with fuel pumped from its own wing by three-phase AC booster pumps mounted two per tank for fail safety, but cross-feed and transfer valves permit fuelling of both engines to be fed from one side, or all the fuel to be used by one engine. Each pump can dry-run for about 15 minutes. In an emergency, the wing tanks can be refuelled by gravity via one over-wing filler point per tank. The tanks are used in the order —centre, inner, outer — with the outer wing tank pumps fitted with sequence valves, so that fuel from these tanks can only be used if there is no supply from the centre or inner tanks. Defuelling is carried out either by the use of the tank booster pumps or through suction. No jettison

BELOW: Diagram showing the standard fuel system.

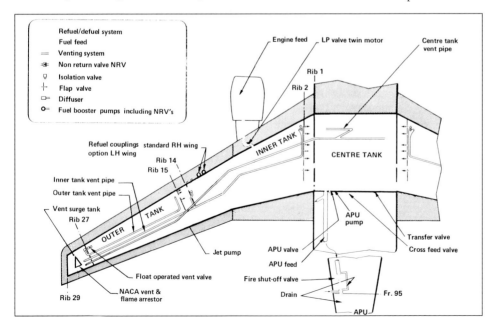

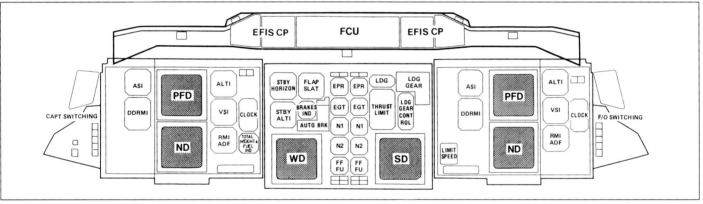

TOP: Flight deck of a South African Airways A300B2.

ABOVE: Diagram showing the instrument panel on the new technology flight deck of A300B4-600.

system is fitted on the Airbus. A surge tank to collect any fuel spilt from the vent pipes during ground or flight manoeuvres is contained in each wing tip.

FLIGHTDECK

The A300 flightdeck was planned and designed in consultation with the airlines, with crew comfort, low workload and a good working space the prime considerations. Central to these aims were separate environmental controls for cool fresh air selection, providing reduced crew fatigue and, therefore, improved operational safety; high level of automation; automatic switching or change-over; 'lights out' concept; and a centralised warning system. From the outset, accommodation was provided for three crew members — captain, first officer and flight engineer — with an additional observer seat on the left hand side. A folding jump seat for occasional use was also fitted. Crew sta-

tions, instrument panels, control column and rudder pedals were designed for optimum seat positions and visibility. The window enclosure, comprising two large front panes, two sliding lateral direct vision panes and two fixed lateral rear vision panes, provides excellent rearward vision. All windows are electrically heated to prevent icing on the front panels and misting up of side panels. The front panels are heated by a conductive gold film, which minimises optical distortion. The forward down vision cut-off is 54ft (16.5m). The avionics compartment below the flightdeck can be accessed via a trap door.

The pilots' instrument panels contain all primary flight and navigational instruments, while the centre panel incorporates those instruments where joint-use is acceptable, such as the primary engine instruments, the master warning panel, brake pressure indicator, and the standby attitude, airspeed and altitude instruments. Flap/slat indicators and landing gear controls are positioned closer to the first officer. The systems panel is located at the side, with the engineer's seat capable of being moved and swivelled to a forward facing central position. This enables the engineer to assist the captain and first officer when necessary, while still being in a position to monitor the systems at his/her own station. In the early 1980s, a few A300B4s were equipped with an interim forward facing two-crew cockpit (FFCC) which, while making greater use of push button techniques, did not have the cathode ray tube instrumentation adopted for the advanced flightdeck of the A300-600 series.

In the two-crew A300-600 flightdeck, Airbus took advantage of Porsche car designers to produce a spacious and ergonomically-styled interior, and of the enormous advances made in avionics technology. All instruments, displays and controls are in front of the pilots, or within easy reach on an overhead panel, which also includes the systems control and monitoring information provided on earlier models on the engineer's console. Flightdeck features include an Electronic Flight Instrument System (EFIS), an Electronic Centralised Aircraft Monitor (ECAM) unique to Airbus, digital avionics and flight management systems, and a windshear warning and guidance system. Six 6.25 x 6.25in (15.9cm) interchangeable cathode ray tube (CRT) displays are used for primary flight, navigation and warning information. The EFIS comprises a primary flight display (PFD) and navigation display (ND) mounted vertically for each pilot, while there are two ECAM displays, warning (WD) and systems (SD), in the lower centre position. Display can be selected to appear on different CRTs.

Installation of an optional Aircraft Communication Addressing and Reporting System (ACARS) allows the crew to manually or automatically transmit maintenance and other messages to the ground by using the control display unit (CDU) on the pedestal. Data can also be made available through the onboard printer, if fitted. The central pedestal layout affords easy access to all crew members. Forward of the throttles are the navigation controls, while alongside the throttles are the controls for the flaps/slats, speed brakes and trim wheels, with the fuel levers and parking brake located just behind. At the rear are the audio controls, VHF communications and automatic direction finder (ADF).

FLIGHT CONTROLS AND GUIDANCE

All primary and secondary flight controls on the A300 are mechanically controlled and powered by three independently supplied hydraulic servo-jacks, with no operational reduction after a single hydraulic failure — electrical control is provided in the later A300-600 version. The primary flight controls comprise inboard and outboard ailerons, trimmable horizontal stabilisers, elevators and rudder, while the secondary flight controls consist of full-span three-section leading edge slats, trailing edge cambered tabless flaps, and spoilers and airbrakes, all of which can be used as lift dumpers. On the A300-600, the outboard low speed ailerons have been deleted

The primary systems provide attitude control in pitch, yaw and roll. Pitch control is achieved by two elevators hinged on the horizontal stabiliser, each actuated by three servo-jacks controlled by a dual mechanical linkage. A duplicated artificial feel system creates load feel at the column, which is variable with flight conditions. It is controlled by two Feel and Limitation Computers (FLCs). An autopilot servo-unit, located at the end of the right hand elevator, comprises two hydraulic actuators operating in parallel, with its control having manual override. A stick-shaker is installed on each control column in the A300-600, to provide stall warning to the crew. Pitch trim is achieved by adjustment of the horizontal stabilisers, either manually by trim wheel operation, or automatically by autopilot trim, mach trim or alpha (angle of attack) trim functions. Electrical and automatic trim signals are processed by two Flight Augmentation Computers (FACs).

For roll control, the A300 has on each wing one all-speed aileron mounted behind the engine and one low speed aileron outboard of the flaps, powered by three servo-jacks. Three spoilers on the top wing surface form, when closed, part of the shroud for the outboard flap. All are operated by the pilots' control wheels through a spoiler servo and mixer assembly. In the case of a control jam occurring in one wing, a spring strut can be compressed to allow lateral control to be maintained in the other. In order to improve aerodynamic characteristics, a droop signal moves the all-speed aileron downwards when the Krüger flaps are extended. The A300-600 differs in having five roll spoilers and no outboard low speed aileron. The roll spoilers are electrically signalled by two identical digital computers, designated Electronic Flight Control Unit (EFCU).

Yaw control is provided by a single-piece rudder, which is operated by three independently-supplied, mechanically-controlled servo-jacks. The rudder receives pilots' input from the pedestal mechanism by a single cable run up to a spring loaded artificial feel unit, connected to the trim screwjack. There are two inner and two outer speed brakes (three outer on the A300-600) located on the upper sur-

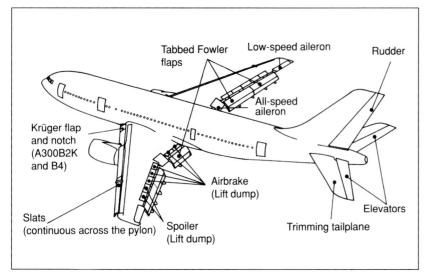

BELOW: Flight control surfaces on the A300.

RIGHT: Close-up of the flap track fairings on the A300-600R. *Günter Endres*

face of each wing, controlled by single servo-jacks, supplied by different hydraulic systems. The outer speed brakes are also used as lift dumpers/roll spoilers and are selected by a lever situated on the pedestal. Speed brake and roll spoiler surfaces are used on the ground as ground spoilers, being automatically extended after touch-down when specific conditions are fulfilled.

The full-span leading edge slats, continuous across the engine pylon, and a Krüger flap and notch, in conjunction with wide-span (84 percent) double-slotted tabbed-Fowler flaps and drooped all speed ailerons, provide an effective high-lift system. The all speed ailerons droop down to 10° to maintain flap continuity in the region of the engine efflux. The Krüger flap and notch are provided to complete the wing leading edge profile when the slats are extended and to obtain better aerodynamic characteristics. The flap and notch are operated by individual hydraulic actuators. Both the slat and flap surfaces are divided into three sections on each wing and have similar but separate operating mechanisms. Each flap and slat surface is driven by two ball screwjacks. Two identical Slat Flaps Control computers (SFCCs) provide continuous monitoring of the high-lift control system.

All models of the A300 are equipped with an Automatic Flight Control System (AFCS). This has been designed to provide cruise guidance with fail soft characteristics, approach guidance with fail passive characteristics to permit Cat II automatic approach and landing and Cat I normal approach, and flight augmentation with fail operational capability by duplication of the system. At the heart of the system is a single Flight Control Computer (FCC) for flight director and autopilot functions, a single Thrust Control Computer (TCC) for speed and thrust control, and two Flight Augmentation Computers (FAC) to provide yaw damping, electric pitch trim and flight envelope monitoring and protection. Options available to extend the range of functions include Cat III automatic landing by the addition of a second FCC; improved availability of speed and thrust functions with a second TCC; full Flight Management System (FMS) through the addition of two Flight Management Computers (FMC) and two control display units; a Windshield Guidance Display (WGD) by adding an optical device in the glare shield to perform ground roll guidance; and a Delayed Flap Approach (DFA) in the TCC to permit a decelerated approach. The FCC takes care of the Flight Director (FD) and the Autopilot in Command Mode (CMD) or in Control Wheel Steering (CWS) mode. Only one autopilot can be engaged at any time, except during the automatic landing and go-around phases. The crew disconnects the autopilot by depressing the instinctive disconnect push button on the control column.

The TCC computes the engine limit parameters, simultaneously processing the maximum operational value (limit) and the recommended value (target). The autothrottle operates in three modes: automatic angle-of-attack protection when the throttle is in the armed phase; a speed (speed or Mach)/thrust mode (N1 or EPR); and a 'thrust latch' mode, which provides full thrust when

selected either manually or automatically. The FMS (optional) provides navigation guidance with performance and fuel management functions, to permit completion of a desired flight plan in an energy-efficient manner within the constraints of air traffic control. The A300 is also fitted with an Aircraft Integrated Data System (AIDS), a digital flight data recording system for performance monitoring, using real-time onboard computer processing. The recorder runs automatically in flight and on the ground when either engine is running and for five minutes after the engines have stopped.

ELECTRICAL AND PNEUMATIC POWER

Power generation is identical on all A300 variants, although better performance is achieved through the use of new-technology components in the A300-600 and A300-600R. Central to this is the replacement of the air-cooled generators coupled to a constant speed drive with compact and lightweight integrated drive generators (IDG).

The A300 has a 115V 400Hz AC system and a 28V DC system. Basic AC power comes from the two 90kVA engine-driven IDGs or a third auxiliary generator driven by the APU. This generator is built with the same electromagnetic components as the IDG, but without the constant speed drive. Any one of the three generators is capable of providing all the AC power up to 90kVA if the other two fail. In the case of power loss from all three generators, a combination of three batteries and a 1,000VA static inverter can supply emergency power for a minimum of 30 minutes, to allow a safe landing with reduced radio and navigation equipment. On the ground, the entire aircraft network can be supplied either from the APU or the ground power unit, with the latter having priority. The primary sources for DC power are three transformer/rectifier units (TRU), and three 25Ah Ni/Cd batteries for emergency supply and APU starting.

The AlliedSignal (Garrett) GTCP 331-250F (TSCP 700-5 from c/n 246) self-contained auxiliary power unit (APU) is installed in the tailcone, aft of the cabin pressure bulkhead. It is designed to provide bleed air to the aircraft pneumatic system and to drive an oil-spray cooled AC generator during ground

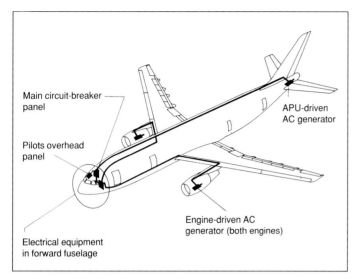

Main circuit-breaker panel

Pilots overhead panel

Electrical equipment in forward fuselage

APU-driven AC generator

Engine-driven AC generator (both engines)

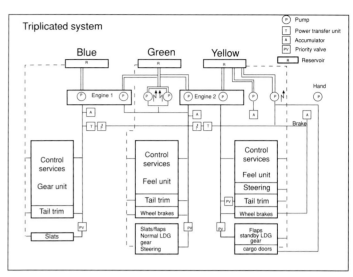

and in flight operations. Independence from external sources on the ground (up to 8,500ft — 2,591m) is assured by power available to drive a 90kVA oil-spray cooled AC generator, and additionally to supply bleed air for main engine start or air conditioning system. During flight operation, the APU is capable of providing either electrical power, or to supply bleed air for emergency wing anti-icing, including air conditioning through one pack, up to 15,000ft (4,575m). It can also provide starter assisted engine re-lights or air conditioning by two packs respectively, up to 20,000ft (6,100m). Maximum operating limit of the APU is 40,000ft (12,200m).

The APU is normally fed through the AC-driven booster pumps from the left hand main engine system, but can be fed from the right when the cross feed valve is open. Where electrical power is unavailable or booster pump pressure is inadequate, a battery-powered pump, located on the rear spar near the aircraft centreline, can raise the fuel pressure to the required level. The APU fuel control system is fully automatic and controlled and monitored by the electronic control box (ECM) installed in the pressurised rear fuselage.

The A300 pneumatics supply high-pressure (HP) air to the air conditioning and pressurisation, wing anti-icing, engine starting, potable water tank pressurisation, reservoirs and the rain repellant system. High-pressure air is supplied from differ-

ABOVE LEFT: Electrical power generation.

ABOVE: Schematic of the hydraulic system.

BELOW LEFT: The APU is located in the tailcone. *Günter Endres*

BELOW: Cabin air supply.

ent stages of the engine's compressors, the APU load compressor, or from two standardised 3in (76mm) ground connections. Two independent systems use bleed air from each engine to supply the thrust reverser and engine intake air ice protection. A third system is installed for the ventilation of the wing leading edge inboard of the engine, using a ram air inlet. No monitoring or manual control is provided.

AIR CONDITIONING AND PRESSURISATION

The air conditioning system serves four independently controlled zones in the aircraft: the flightdeck and three passenger cabins, with the forward and rear areas including galleys and toilets. Airflow routing also provides ventilation of the avionics compartment and the lower cargo holds. In the A300-600 and A300-600R, air is supplied to the toilets and galleys from an individual ventilation system, with cabin pressure differential

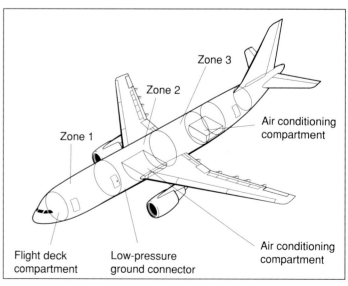

Zone 3

Zone 2

Zone 1

Air conditioning compartment

Flight deck compartment

Low-pressure ground connector

Air conditioning compartment

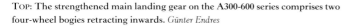

Top: The strengthened main landing gear on the A300-600 series comprises two four-wheel bogies retracting inwards. *Günter Endres*

Above: Lateral bracing of main gear. *Günter Endres*

Above Right: The twin-wheel nose gear retracts forward into the fuselage. *Günter Endres*

providing extraction during the flight, while a fan is used when the aircraft is on the ground. Hot air is tapped downstream of the bleed air control valves and supplied via a pneumatics distribution manifold to two air conditioning packs, located under the central wing box. Each pack incorporates a three-wheel 'bootstrap' air cycle machine with an air-to-air heat exchanger, where the air is cooled and routed, via a common manifold, to the four zones. Temperature control is achieved automatically or manually by varying the pack outlet temperature and adding trim (hot) air. The air supply can also be taken from the APU, and conditioned air can be supplied directly to the cabin air distribution system by two low-pressure ground connections. A ram air inlet is provided for fresh air ventilation in flight when the packs are not operating.

Pressurised areas in the A300 are the flightdeck, passenger cabin, avionics compartment and cargo holds. Control is provided by two electric outflow valves operated by two independent automatic cabin pressure systems, one active and the other stand-by. Switch-over from one to the other is automatic after each flight, and in the case of failure of the active system. Manual control of the outflow valves is possible by switches in the overhead panel. Ground depressurisation is achieved automatically by electrically opening the outflow valves. Automatic pre-pressurisation of the cabin before take-off is provided to prevent a noticeable pressure fluctuation in the cabin during rotation on take-off.

HYDRAULIC SYSTEM

The hydraulic system operates the fully powered flight controls, landing gear and braking system, and the cargo compartment doors. It comprises three parallel, fully independent systems, each pressurised by at least two independent means, to ensure aircraft control in the event of the loss of two systems or both engines. Each includes a reservoir and is pressurised to 207 bars (3,000 psi). The main power generation consists of identical variable displacement pumps driven by the engine accessory gearboxes. Auxiliary power is provided by two electric pumps for use mainly on the ground for maintenance and check list purposes, two power transfer units in the event of failure of the LH or RH engines or on the ground, a pump driven by ram air turbine should both engines fail, and an electric pump pressurising the braking accumulator and allowing cargo compartment door operation on the ground. A hand pump is

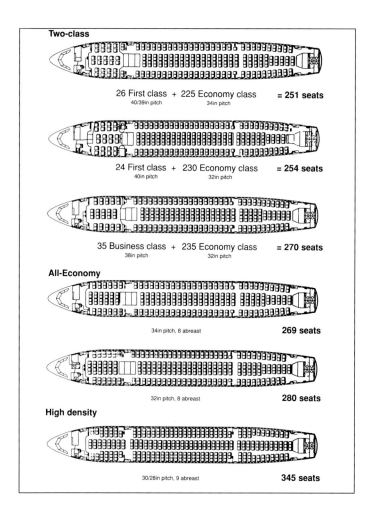

Two-class

26 First class + 225 Economy class = **251 seats**
40/39in pitch · 34in pitch

24 First class + 230 Economy class = **254 seats**
40in pitch · 32in pitch

35 Business class + 235 Economy class = **270 seats**
38in pitch · 32in pitch

All-Economy

34in pitch, 8 abreast **269 seats**

32in pitch, 8 abreast **280 seats**

High density

30/28in pitch, 9 abreast **345 seats**

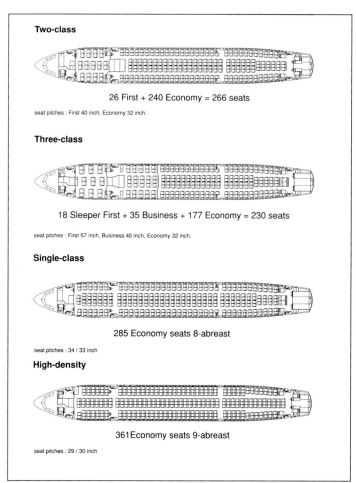

Two-class

26 First + 240 Economy = 266 seats

seat pitches : First 40 inch, Economy 32 inch.

Three-class

18 Sleeper First + 35 Business + 177 Economy = 230 seats

seat pitches : First 57 inch, Business 40 inch, Economy 32 inch.

Single-class

285 Economy seats 8-abreast

seat pitches : 34 / 33 inch

High-density

361 Economy seats 9-abreast

seat pitches : 29 / 30 inch

also available for emergency operation of the cargo doors. The flight control servos are protected against high flow consumers by priority valves.

LANDING GEAR

The landing gear on all A300 models is of conventional hydraulically actuated tricycle-type, with two four-wheel tandem mounted bogies on the main unit and twin-wheels on the nose gear. The main gears are located under the wing and retract inwards towards the fuselage centreline, while the nose-wheel unit retracts forward into the fuselage. In the extended position, the main gear is braced laterally by foldable side struts, and the nose gear by a telescopic drag strut. The main gear on the A300-600 is additionally reinforced, with a new-design hinge arm. The nose gear steering system is hydraulically powered with mechanical control.

A solenoid operated anti-retraction latch is installed to prevent the normal gear control lever being moved to the up position, unless shock absorber struts are fully extended and the nose wheels and main bogie beams are perpendicular to the gear leg. Emergency extension of the nose and main gears is possible by manual release of the uplock mechanism and free fall extension and locking of the gears, with automatic spring assistance for the main gear and aerodynamic load for the nose gear. Visual and aural indicating and warning systems alert the crew when the landing gear is not locked down during approach. A mechanical alternative means of checking that the gear is locked-down is provided by a red pin protruding on

each wing upper skin above the gear well, which is visible from the cabin. Nose gear down-lock can be ascertained through a viewer located in the avionics compartment, even in the event of adverse environmental conditions, such as misting, frosting or water accumulation.

The four wheels of each main landing gear are equipped with hydraulically operated disc brakes, each with an anti-skid system based on slip ratio control. This permanently compares the braked wheel speed with the actual aircraft speed, in order to control the braked wheel speed to a pre-determined slip ratio. Brake temperature indication and overheat warning is provided on the flightdeck. Protection against tyre skidding and locked-wheel touchdown is also built in. Automatic braking and the installation of brake cooling fans are optional. Normal braking pressure is controlled by one master valve per main landing gear, and supplied to the eight brakes via eight anti-skid servo valves. Emergency braking is supplied by the second hydraulic system via an accumulator, but anti-skid protection is not available.

CABIN INTERIORS AND CARGO HOLDS

The 18ft 6in (5.64m) fuselage cross-section provides an optimum balance between aerodynamic considerations, true wide-body passenger accommodation, and a standard underfloor cargo systems fit. It allows customised twin-aisle layouts, ranging from a six-abreast first class configuration to nine-abreast high-density seating for the charter market. A typical mixed-

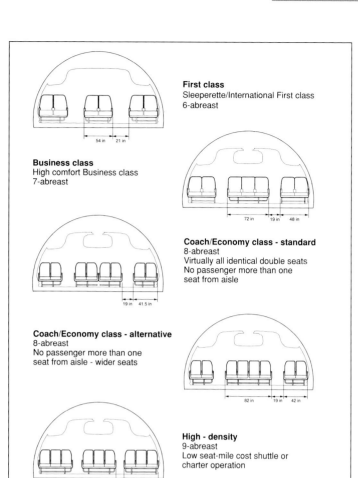

First class
Sleeperette/International First class
6-abreast

54 in 21 in

Business class
High comfort Business class
7-abreast

72 in 19 in 48 in

Coach/Economy class - standard
8-abreast
Virtually all identical double seats
No passenger more than one
seat from aisle

19 in 41.5 in

Coach/Economy class - alternative
8-abreast
No passenger more than one
seat from aisle - wider seats

82 in 19 in 42 in

High - density
9-abreast
Low seat-mile cost shuttle or
charter operation

16.5 in 57.1 in

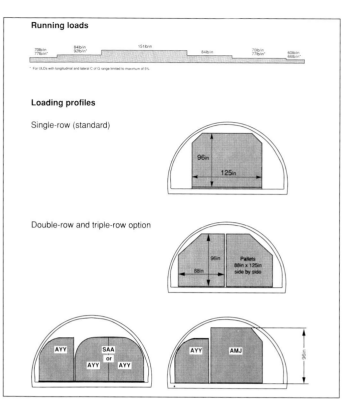

Running loads

70lb/in   84lb/in   151lb/in   84lb/in   70lb/in   60lb/in
77lb/in* 92lb/in*   77lb/in* 66lb/in*

* For ULDs with longitudinal and lateral C of G range limited to maximum of 5%.

Loading profiles

Single-row (standard)

96in
125in

Double-row and triple-row option

96in   Pallets
88in   88in x 125in
       side by side

AYY   SAA   AYY   AMJ
      or
AYY   AYY

96in

ABOVE: Cabin cross-section of freighter showing how variously shaped cargo can be accommodated.

LEFT: Cabin cross-sections with typical passenger arrangement.

BELOW: Loading cargo pallets onto a FedEx A300-600R.

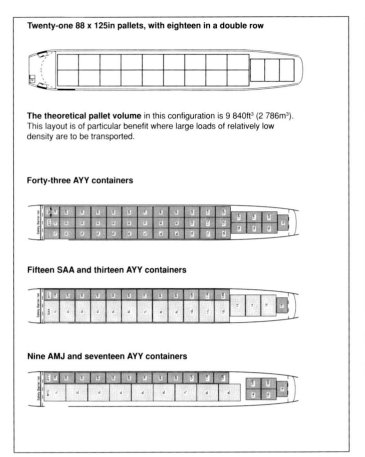

Twenty-one 88 x 125in pallets, with eighteen in a double row

The theoretical pallet volume in this configuration is 9 840ft³ (2 786m³). This layout is of particular benefit where large loads of relatively low density are to be transported.

Forty-three AYY containers

Fifteen SAA and thirteen AYY containers

Nine AMJ and seventeen AYY containers

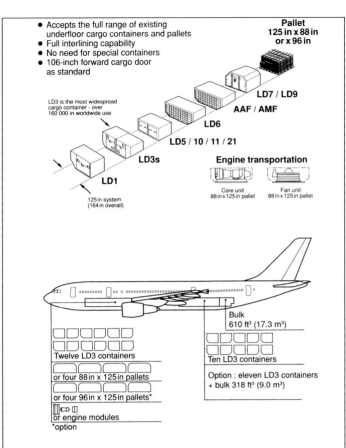

- Accepts the full range of existing underfloor cargo containers and pallets
- Full interlining capability
- No need for special containers
- 106-inch forward cargo door as standard

Pallet
125 in x 88 in
or x 96 in

LD3 is the most widespread cargo container - over 160 000 in worldwide use

LD7 / LD9
AAF / AMF
LD6
LD5 / 10 / 11 / 21
LD3s
LD1

125in system
(164 in overall)

Engine transportation

Core unit
88 in x 125 in pallet

Fan unit
88 in x 125 in pallet

Bulk
610 ft³ (17.3 m³)

Twelve LD3 containers

or four 88 in x 125 in pallets

or four 96 in x 125 in pallets*

or engine modules
*option

Ten LD3 containers

Option : eleven LD3 containers + bulk 318 ft³ (9.0 m³)

ABOVE LEFT: Main deck cargo layouts.

ABOVE: Lower cargo hold flexibility.

class layout in the A300B2/B4 accommodates 35 business class passengers in a 2/3/2 configuration with a 38in seat pitch, plus 235 economy class passengers 2/4/2 at 32in (813mm) pitch, narrowing to 2/3/2 in the rear fuselage taper. High-density nine-abreast seating accommodates 345 passengers (361 in the A300-600). Some 95% of passenger seats are double-seat units with a width of 40.4in (1,026mm). There is a gap between the two centre units to ensure that each passenger has at least one armrest not shared with another passenger. The two aisles are 19in (483mm) wide. Nine attendant seats are also incorporated, one at each door, and a purser station in the front passenger cabin. Two optional attendant seats can be installed at the front side of the forward cabin bulkhead.

The cabin has been designed for four galleys, with a total capacity of 21 trolleys in the A300B2/B4 and 24 trolleys in the A300-600/600R, allowing the serving of two hot meals plus beverages. The standard aircraft is equipped with five lavatories, four in the rear and one in the forward cabin. Structural provisions have been made for an additional lavatory in the front cabin. Access to the cabin is via three large doors along each side of the fuselage, two ahead of the wing and one at the rear, measuring 76 x 42in (1.93 x 1.07m). These are outward parallel-opening plug-type doors and require no power assistance. There are also 63 x 25in (1.60 x 0.61m) plug-type emergency exits of similar construction on both sides behind the wing.

All models of the A300 have two underfloor cargo compartments, which can accept all standard unit load devices (ULD) in current use. In the forward hold, up to 12 LD3 containers can be carried two-abreast, while four full-size pallets or a mixture of both can be alternatively accommodated. The rear hold

can take up to eight LD3 containers and bulk cargo in the A300B2/B4, and 10 LD3 containers or 11 LD3 containers plus bulk cargo in the A300-600. A net separates the rear cargo from the bulk cargo compartment, which has a usable volume of around 610 cu.ft (17.3m³) and a load capacity of 2,770kg (6,100lb). All holds can be heated and ventilated for the carriage of livestock. A 106in wide x 67in high (2.69 x 1.70m) forward lower cargo door on the right side became a standard item from aircraft No.157. The rear hold door measures 71in wide by 67in high (1.81 x1.7m). Both doors open outwards and upwards hydraulically, with manual locking/unlocking, and extend over the full height of the holds. There is also a manually-operated 37 x 37in (0.95 x 0.95m) door to the bulk cargo compartment. The A300-600C convertible and -600F all-freighter modes, as well as the A300B2/B4 freighter conversions, have a main deck cargo door on the forward port side with 70° or 145° opening angle and measuring 141in wide by 101in high (3.58 x 2.57m). Typical cargo on the main deck of the A300-600F comprises nine 88 or 96 x 125in (223.5 or 243.8cm x 317.5cm) and six 88 x 125in (223.5 x 317.5cm) pallets, while four pallets and 10 LD3 containers, or 22 LD3s, can be carried in the lower hold.

Both the forward and aft cargo compartments in the A300-600 are provided with a semi-automatic electrically powered loading system, which is controlled by an operator from the control panel, located behind service doors in the outer skin on

the right hand side of each doorway. The system permits manual loading/unloading if the power drives are inoperative.

PROTECTIVE SYSTEMS AND EMERGENCY EQUIPMENT

A fire protection system provides immediate detection of a fire in the engine nacelles and auxiliary power unit (APU) compartments through aural and visual indications, together with extinguishing components. The electronic bay is provided with a smoke detection system, while the cargo compartments comprise both smoke detection and fire extinguishing. Additionally, portable fire extinguishers are installed in the avionics bay, on the flightdeck and in the passenger cabin.

The detection system consists of two parallel sensing loops in each protected area, both of which must be subjected simultaneously to fire or overheat conditions before triggering the alarm, thus reducing the risk of false warnings. The fire extinguishing system for each nacelle consists of two extinguisher bottles installed at the rear part of the pylon, while one bottle is provided for the APU. Smoke in the avionics bay is detected by four self-contained smoke detectors of the ionisation type, with two detectors provided per cargo compartment. For fire extinguishing in the cargo holds, two bottles (one on the A300B2),

ABOVE LEFT: The crew galley for the Airbus freighter is located behind the cockpit.

BELOW: Economy class layout in an American Airlines A300B4-605R.

ABOVE: Business class service in an A300B4-603 of Lufthansa. *Ingrid Friedl*

each equipped with two independent discharge cartridges, are installed in the forward cargo compartment in the A300-600, with one each in the forward and mid compartment in the A300B4.

The A300 is also provided with two independent emergency oxygen systems — one for the flight crew and one for cabin staff and passengers — designed for three flight profiles, including a one minute delay at maximum altitude, nine minutes descent to 10,000ft (3,050m), and 65 minutes continuation of flight at or below 10,000ft (3,050m) for the crew. The gaseous low-pressure system for the crew includes quick-donning masks with mask-mounted regulators. For passengers, a solid-state system, which is a drop-out continuous flow system installed in the overhead racks and including oro-nasal masks, has been chosen for ease of installation, safety and reduced weight.

An emergency evacuation signalling system enables the flight or cabin crew to warn each other and the passengers by warning lights and an audio tone, and can be operated from a panel on the flightdeck, at the purser's station and from the C/A station aft of the rear left hand door. Other emergency equipment distributed through the cabin includes six fire extinguishers, two axes, two megaphones, six flash lights, two radio beacons, four therapeutic portable oxygen bottles and masks, and life jackets sufficient for passengers, crew and demonstration purposes. All emergency egress is through the cabin doors, but the flight crew also has the option of using the 28 x 24in (0.7 x 0.6m) sliding windows, for which escape ropes are provided. The forward and rear passenger doors are equipped with double-track escape slides, while the mid-cabin emergency exits have single-track slides. On overwater equipped aircraft, slides on doors one, two and four are replaced by a slide/raft combination, configured to function as slides in a ground evacuation, and as rafts when detached from the aircraft following a ditching. Emergency lighting is provided for the passenger areas and the external surrounds of the exits, powered from the 28V DC supply and/or the 6V emergency Ni/Cd batteries.

Ice and rain protection for critical areas of the aircraft permits unrestricted operation in icing conditions and heavy rain. Ice protection is achieved by hot air or electrical heating, and includes the outboard part of the leading edge slats of each wing totalling some 47.5 percent of the aircraft span, engine air intakes, the engine itself (Pratt & Whitney only), the front windshield panels and side windows in the cockpit, sensors, pitot probes and static ports, and the waste water drain masts. Ice protection bleed air for the wing is taken from the engine or APU via the crossfeed duct and controlled by two shut-off valves and restrictors in each wing.

Engine anti-ice is provided for the P&W engine only by hot air bled from the eighth stage compressor to the first stage stator vanes. For both P&W and GE engine nacelles, each nacelle is provided with an intake anti-ice system. Rain removal from the front windshield panels is by wipers and, when necessary, by a rain repellant fluid system.

RIGHT: Typical galley arrangement.

BELOW: Arrangement of the emergency chutes.

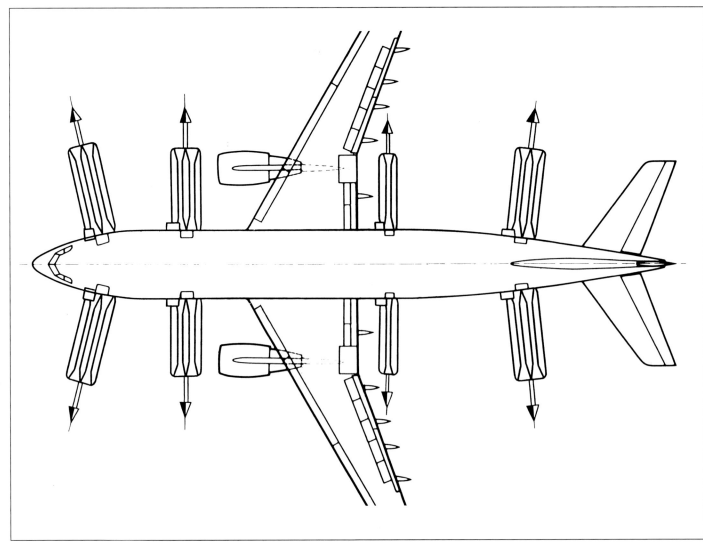

4 IN SERVICE

From a commercial standpoint, sales started slowly and remained so right up to certification. Only three airlines had placed firm orders by March 1974, including launch customer Air France, Iberia and Lufthansa. Letters of intent had been received from Sterling Airways for three A300B4s on 4 May 1972, Swiss charter airline SATA for one A300B4 on 9 July 1973, and from Transbrasil for two A300B2s on December that same year. None of the three went on to order or operate the Airbus.

Several factors combined to create and perpetuate this situation. A stagnation of traffic and the oil crisis had left airline balance sheets in a precarious position, forcing many to hold back on making large financial commitments. In addition, airlines were unsure about buying European equipment, with particular concern about product support, and preferred waiting to see how the aircraft and manufacturer performed in the first few months of service. The claimed lower fuel consumption was not a powerful enough weapon to offset the effects of the oil shortage. The weakening of the US dollar against the European currencies also did Airbus Industrie no favours, since the sale price of the aircraft had to be maintained in dollars, consequently making the aircraft more expensive and resulting in heavy losses for the manufacturers.

MARKET PREDICTIONS FOR WIDE-BODY TWINS TO 1985			
Market area	Number to 1975	to 1980	to 1985
European scheduled	23	199	291
European charter	9	85	128
United States trunks	-	313	457
Rest of North America	10	143	228
Rest of the World	5	140	227
TOTAL	47	880	1,331

Source: Airbus Industrie 1973

As a means of extending the market appeal, Airbus revived the option of offering the A300B with the 48,000lb (214kN) thrust RB211-524, but both Rolls-Royce and the British government baulked at the suggestion that they would have to contribute financially and the talks fizzled out again. Also on the table were the A300C convertible and A300F all-freighter versions of the B4, both fitted with a 23ft 2in x 8ft 4in (7.07

x 2.54m) upwards-opening cargo door on the port side and capable of carrying a 40 tonnes payload a distance of 1,000nm (1,850km). Only three and two respectively were later built on the production line, although there were some subsequent conversions. The German Air Force was offered a military version with a large freight door to replace its Transalls on casualty evacuation, air refuelling, freighting and trooping, but showed little interest, with the armed forces of France, Italy and the UK proving even less enthusiastic. Such an aircraft would, in any case, have required substantial modifications to provide a rough-field capability and the idea was quietly dropped.

FIRST SERVICE

Launch customer Air France took delivery of its first A300B2 on 10 May 1974 and on 23 May put the type onto the busy Paris–London route, flying two return trips on the day. After the delivery of the second and third aircraft a few weeks later, the French flag-carrier introduced the new type also on the

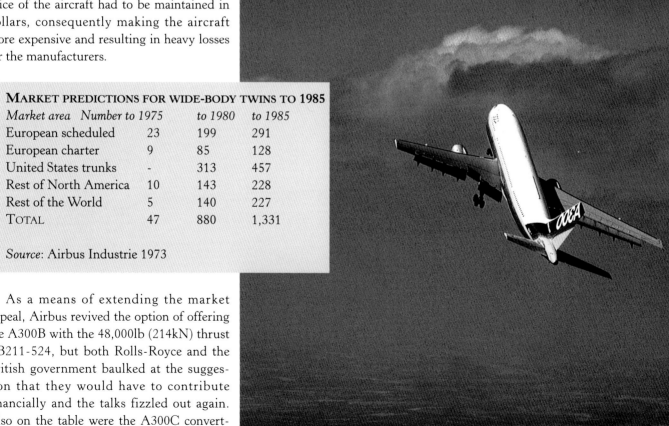

RIGHT: A300 climbs steeply out of Toulouse.

schedules to Düsseldorf, Nice and Milan. Air France had already been involved in route proving flights between 14 and 21 January as part of the certification process, when the Airbus recorded 51hr 55min flying between Toulouse, Paris Orly, Frankfurt, Algiers, Marseille and Nice, a total of 18,000 miles (29,000km). The only other deliveries in 1974 were to charter operator Trans European Airways, which leased aircraft No.2, the only A300B1 to enter airline service, and Air Siam, which had placed an order for two on 31 July. Korean Air Lines became another Far Eastern customer, signing up for four A300B4s on 5 September that year. The year ended with the first flight of the A300B4 on 26 December. The A300B4

TOP: Korean Air became the second Airbus operator in the Far East when it placed an order for six A300B4s in September 1974.

ABOVE: South African Airways (SAA) was the first customer of the A300B2-3K, which adopted the Krüger flaps of the B4 model.

demonstrated its full-payload range-capability on 15 March 1975, flying Geneva–Kuwait–Toulouse, a total distance of 5,186nm (9,595km). It received its type certificate on 26 March and entered service with Germanair in May.

The first year of airline service was unusually trouble-free with no notable problems reported by the operators. Sales were, however, still hard to come by, and only 13 firm A300B2 orders

Top: This A300B4-203 was one of four leased to Eastern Air Lines as a means to access the vast US market.

Above: American Airlines was the launch customer for the extended-range A300B4-600R — named LuxuryLiner by the US carrier.

were signed during 1975 by three new customer airlines, Air Inter, Indian Airlines, and South African Airways (SAA), the first African customer for Airbus. SAA specified the addition of the B4's Krüger flaps to improve hot-and-high performance, which produced the A300B2K, with the South African flag-carrier becoming the launch customer for that variant on 4 September 1975. These orders provided a little relief for

Airbus, but the trickle did not turn into a flood. Quite the contrary, as the following year passed without further additions to the order book.

The all-important launch order from a United States airline still proved elusive, although Airbus was working hard to secure a first order from Western Airlines. The Los Angeles-based carrier, one of the US pioneers, was thought to favour an initial order for five A300B4s, with more to follow. Against all expectations, Western decided to stick with Boeing, announcing its decision not to buy the A300 at the end of January 1977. Although this constituted a disappointing setback for the Europeans and forced Airbus to reduce the production rate

the eastern half of the United States and Central America and the Caribbean. Eastern needed a modern aircraft to fit in size between its Lockheed TriStars and Boeing 727s, and capable of operating economically on its shorter sectors on the US East Coast. But there was a major snag. The airline was in a parlous financial state and in no position to pay the asking price for a new Airbus fleet. But with Airbus desperate to break into the American market, and Eastern equally desperate to add new aircraft to bolster its operation, a unique solution was found. Airbus Industrie made four A300Bs available to Eastern for a six-month trial at no charge, except for cabin furnishings. At the end of the period, the agreement was that Eastern could either purchase the aircraft, or return them if they proved unsatisfactory. The four aircraft were handed over in the latter half of 1977 and, after route proving and trial flights, officially entered service with the winter schedule on 13 December. The successful integration of the A300 Whisperliner into its operations led to a much-prized firm

temporarily from 2 to 1 per month from April, there was plenty of good news to come during 1977. The order drought, which had caused great concern among the Airbus partners, ended that year, when eight airlines placed firm orders for 20 aircraft. The 100,000 flight hours mark was reached on 29 September, on which date 34 aircraft had been delivered to 10 airlines.

order for 23 aircraft, plus nine options from the US major on 26 June 1978. The breakthrough into the US market had finally been made.

The US manufacturers, and especially Boeing, had been closely following progress in Europe. In spring 1976, Boeing proposed a joint airliner, based on a Boeing wing and A300 fuselage, but the European partners were not at all sure of the motives that lay behind this unexpected initiative. Although talks on what became known as the BB10 project continued for some time, both in Seattle and Toulouse, suspicions remained and Airbus eventually pulled out. Boeing tried its luck with

AMERICA AT LAST

Western Airlines had been only one of the US majors targeted by Airbus, and the sales team now redoubled its efforts, homing in on Eastern Air Lines, one of the 'Big Four' and serving

British Aerospace, offering a risk-sharing partnership on its new 757 narrow-body programme, but was equally unsuccessful, although it later sold a large number of 757s to British Airways.

With Douglas cancelling its proposed 200-seat DC-10 Twin that same year and

Top: An early A300 was used to provide 25-second periods of weightlessness for use in the training of astronauts and in preparing equipment for space missions. The programme involved Novespace, French space agency CNES and Sogerma.

Left: The third Airbus prototype used to trial the fly-by-wire system during 1986.

Above Right: First A300B4-600 on its maiden flight on 8 July 1983.

Right: A300B4-600 production sharing, showing who makes which parts before final assembly at Toulouse.

Airbus models are now flying with Rolls-Royce engines.

Over the 10 years since its withdrawal from Airbus, Britain's rejoining had always remained under active discussion, and the Boeing overtures concentrated the mind on whether its industry would benefit more from a European partnership or from being merely subcontractors to the Americans. The prospect of building the wing for the new A310 and becoming involved in future designs turned the scales towards Airbus, but a final decision was conditional on Airbus winning at least one order from a British airline. Negotiations with Laker Airways offered the most promise, and on 18 August 1978, the Airbus partners and British Aerospace (BAe) initialled an agreement for BAe's entry as a full partner. BAe had been established on 29 April 1977, when nationalisation of the British aircraft industry brought together the British Aircraft Corporation (BAC), Hawker Siddeley Aviation, Hawker Siddeley Dynamics and Scottish Aviation under a single umbrella. The full partnership agreement was officially signed on 28 November 1978, and Britain returned to the fold with effect from 1 January 1979, taking a 20 percent stake. As a result, the Aerospatiale and Deutsche Airbus shares were reduced to 37.9 percent each. Laker Airways signed a contract for 10 A300B4-200 on 10 April 1979 and operated the type successfully on its holiday routes until its liquidation in February 1982.

Boeing's plans for the 7X7 going nowhere fast, the market was left wide open for Airbus to launch the smaller A310 (described fully in a separate book), formerly known as the A300B10, alongside the A300B2/B4 models. A signature agreement with Pratt & Whitney to equip the Airbus with the JT9D-59A engine, further extended the Airbus options and market appeal. To bring some semblance of order to the growing number of variants and permutations of weights and engine combinations, Airbus Industrie devised a new designation system, which took effect from 1978. Standard A300B2 and B4 models were given the -100 suffix, while aircraft with higher gross weights took on the -200 suffix. The two engine types also had their own groupings, so that aircraft numbered from -100 to -119 and -200 to -219 had CF6-50 engines, and those numbered from -120/220 to -139/239 were powered by the JT9D. Within these ranges, certain individual numbers also identified the engine subtype, for example, 101/201 referred to the CF6-50C, -102/202 to the CF6-50C1, and -103/203 to the short-nozzle CF6-50C2. Airbus Industrie had never given up hope that a customer might specify Rolls-Royce engines and the suffixes -140/240 to -159/259 were reserved for the RB211. These were never used on the A300, but later

A300-600 production sharing

- Aerospace
- DaimlerChrysler Aerospace
- British Aerospace Airbus
- CASA
- Fokker
- GE, PW
- Messier

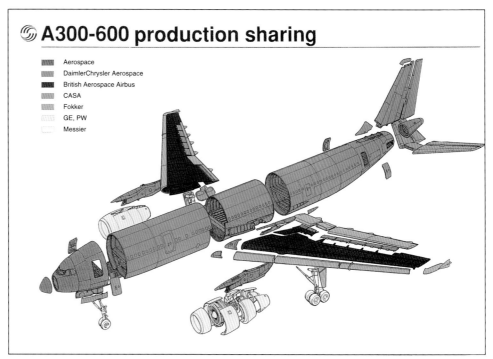

ADVANCED TECHNOLOGY
A300-600

Hand-in-hand with the development of the advanced shortened A310 went the A300-600 development of the B4, which was launched on 16 December 1980 with an order from Saudia for 11 aircraft powered by Pratt & Whitney engines. Billed by Airbus as the world's largest twin-aisle aircraft, the A300B4-600, generally simplified to A300-600, incorporated many new design features, which added up to producing an aircraft with a productivity potential of nearly three times that of the original B2 model. Centre-piece was a new two-crew cockpit with digital avionics and flight management systems, a unique electronic centralised aircraft monitor (ECAM) providing extensive aircraft systems information, reminder and warning displays, allied to a pleasant crew working environment created by luxury German car manufacturer Porsche. Other notable improvements were a substantially modified wing, including a new inner wing section, electrical signalling for flaps, slats and spoilers, wingtip fences for drag reduction, and a saving of 1.5 tonnes in weight through the use of composite materials. The entire vertical stabiliser was built of carbon fibre composites, alone producing a weight saving of 22%, or 256lb (115kg), and a 95% reduction in component parts. The A300-600 was also fitted with carbon brakes, which tripled capacity to 3,000 landings, radial tyres, and a more fuel-efficient APU. It was also offered with a choice of engines, including several variants of the General Electric CF6-80C2 engine and the Pratt & Whitney JT9D (later the PW4000 series). The new aircraft provided typical two-class seating for 266 passengers in a slightly longer fuselage, or up to 361 in an all-tourist layout, with an underfloor cargo capacity of 23 LD3 containers and four pallets.

The A300-600 made its first flight on 8 July 1983 and received its type approval from the French and German authorities on 9 March 1984. 16 days later, Saudia took delivery of its first aircraft, powered by the 56,000lb (249.2kN) thrust Pratt & Whitney JT9D-7R4H1 turbofan, and putting the A300-600 into service at the beginning of June. Next to fly, on 12 April 1984, was the convertible A300C4-600, which had been launched by Kuwait Airways on 30 June 1981. Equipped with a large forward port-side cargo door and capable of being operated in all-passenger or all-freight modes with a maximum structural payload of 46.8 tonnes, the first C variant was delivered to the Middle East carrier on 30 May 1984. Thai Airways International was the first customer for the General Electric-powered passenger aircraft, which first flew on 20 March 1985 and was certificated a few days prior to delivery to Bangkok on 26 September.

The A300-600 replaced previous models after the last deliveries of the A300B4 were made in early 1985, but the process of product improvement was far from complete. On 2 March 1987, American Airlines launched the extended-range A300B4-600, first referred to as the A300B4-600ER but soon shortened to A300-600R, with a massive order for 25 aircraft, eventually increased to 35 units. The -600R differed from the basic -600 essentially by the installation of a 1,620 US gallon

Top: First flight of A300B4 converted to freighter configuration by British Aerospace Aviation Services.

Above: UK cargo operator Channel Express was the launch customer for the A300B2/B4 freighter conversions.

Above Right: Cargo door of freighter conversion by British Aerospace.

Far Right: Main deck cargo loading system of a A300B4-203 passenger-to-freighter conversion.

(6,150 litre) fuel tank in the horizontal stabiliser, with a computerised fuel transfer system for active centre of gravity control, and the required equipment fit for 180-minute ETOPS (extended-range twin-engine operations) approval for all airframe/engine combinations. The A300-600R first flew on 9 December 1987 and entered service with American Airlines the following May, after receiving type approval on 10 March 1988. The next day, Korean Air ordered a Pratt & Whitney PW4000-powered A300-600R, which made its maiden flight on 20 September 1988 and entered service two months later.

FREIGHTER EXPRESS

Apart from a handful of orders, cargo versions of the A300 attracted little interest from airlines, possibly due to the availability of the larger Boeing 747F and a glut of earlier generation passenger aircraft, which had been converted for cargo use. Belly-hold cargo capacity had also increased immensely with the arrival of the new wide-bodies. But all this was to change in summer 1991, when the giant US package operator Federal Express signed a huge order with Airbus for 25 all-cargo A300F4-600Rs, later increased to 35. This order confirmed what many within Airbus Industrie had believed for a long time, that the A300 would make an excellent cargo aircraft. With its 5.64m (18ft 6in) diameter, the dedicated A300-600F can accommodate a single or double row of standard pallets,

associated systems and equipment have also been deleted. One lavatory and two attendant seats are installed forward of the left entrance door, and a stowage locker and four courier seats forward of the right door. In the flight compartment, in addition to the two pilots seats, a folding observer seat is provided with an option for a second. The A300-600F took to the skies on 2 December 1993, with the first delivery to Federal Express taking place on 27 April 1994.

The introduction of the vastly improved A300-600/600R models had resulted in many airlines adding to or replacing the earlier A300 B4-100 and -200 versions in their passenger fleets. Allied to the confidence in the Airbus' cargo capability exhibited by Federal Express and a general buoyancy in the cargo sector, which is projected to grow by 8% annually over the next 10 years, these factors combined to create a market for a freighter conversion programme involving older models. In response, Daimler-Benz (now DaimlerChrysler) Aerospace Airbus (Dasa) developed a conversion kit, which features the same large port side cargo door, replacement of passenger door No 2 by a shell panel, reinforcement of the main deck floor to increase running loads, Class E fire protection and smoke

plus four pallets and 10 LD3 containers, or 22 LD3s, on the lower deck. It is also certificated for Cat IIIb autoland and is well within FAR 36 Stage 3 and ICAO Annex 6 Chapter 3 noise requirements.

Structural changes from the passenger aircraft include the addition of a large 141 x 101in (3.58 x 2.57m) main deck cargo door on the port side (as fitted to the convertible C models); reinforced main deck floor; enhanced fire protection on both floors; deletion of passenger doors numbers 2, 3 and 4 on both sides; deletion of cabin windows except where required for maintenance; and the addition of a semi-automatic, electrically-powered cargo loading system on both decks and a 9g safety barrier net and smoke curatin on the main deck. All passenger-

detection on the main deck, safety barrier net and smoke curtain and systems adaptation/simplifications for the freighter role. Optional packages include cargo loading systems, enhanced payload capability, revised interior linings and weight saving proposals.

British Aerospace Aviation Services, based at Bristol-Filton Airport, developed its own supplemental type certificate (STC), while Aerospatiale Group member Sogerma carries out conversions at Bordeaux and Toulouse on the Dasa STC, as does Dasa company Elbe Flugzeugwerke in Dresden. The first converted aircraft was delivered by BAe to launch customer Channel Express in July 1997. Both Dasa and BAe are working on STCs for the conversion of the latter A300-600 models.

THE WHITE WHALE

With production rates increasing and new larger models being added to the Airbus family, the combination of the four ageing Super Guppy aircraft and road transport was now proving inadequate to move aircraft parts quickly enough between the different manufacturing locations. Airbus reviewed all the options available by using existing aircraft and even considered sea transport for certain elements, but none met its requirements. The huge Russian Antonov An-124 inevitably entered the calculation, but was suitable only for carrying the wings, while the DC-10 was eliminated, as the third engine would have restricted access to the cargo bay. Also discarded were some more fanciful ideas of using a piggy-back arrangement with the wings on top. A conclusion was quickly reached that a new aircraft based on the A300-600R was best suited to replace the Guppy.

LEFT: Two freighting giants — the A300F4-605R and A300-608ST Beluga Super Transporter.

RIGHT: The A300-608ST undergoing wind-tunnel tests at Deutsche Airbus.

A300B2/B4 FREIGHTER CONVERSIONS (to 1 January 1999)

C/n	Model	Operator	BAe	Elbe	Sogerma
0019	B4-203F	MNG Airlines (Turkey)	*		
0023	B4-203F	ACS Cargo (Costa Rica)	*		
0041	B4-103F	Farnair Europe (Switzerland)/DHL		*	
0042	B4-103F	Farnair Europe/DHL		*	
0044	B4-103F	Farnair Europe/Tulip Air		*	
0045	B4-203F	HeavyLift Cargo Airlines (UK)	*		
0047	B4-203F	HeavyLift Cargo Airlines (UK)	*		
0074	B4-203F	HeavyLift Cargo Airlines (UK)	*		
0078	B4-203F	International Cargo Charter Canada	*		
0083	C4-203F	MNG Airlines (Turkey)		*	
0095	B4-203F	EAT (Belgium)/DHL			*
0100	B4-203F	?	*		
0101	B4-203F	JHM Cargo Express (Costa Rica)	*		
0106	B4-203F	JHM Cargo Express (Costa Rica)	*		
0107	B4-203F	JHM Cargo Express (Costa Rica)	*		
0116	B4-203F	EAT (Belgium) Hunting Cargo/DHL		*	
0117	B4-203F	Channel Express (UK)/TNT		*	
0121	B4-203F	Channel Express (UK)		*	
0123	B4-203F	Jet Link International (Netherlands)	*		
0124	B4-103F	Channel Express (UK)	*		
0139	B4-203F	JHM Cargo Express (Costa Rica)	*		
0140	B4-203F	Mexicana de Carga (Mexico)	*		
0142	B4-203F	JHM Cargo Express (Costa Rica)	*		
0154	B4-203F	HeavyLift Cargo (UK)	*		
0173	B4-203F	JHM Cargo Express (Costa Rica)	*		
0183	B4-203F	?	*		
0199	B4-203F	EAT (Belgium) Hunting Cargo/DHL			*
0250	B4-203F	EAT (Belgium) Hunting Cargo/DHL			*
0289	B4-203F	EAT (Belgium) Hunting Cargo/DHL			*
TOTAL			18	7	4

Preliminary feasibility studies were, therefore, initiated in 1989 and the report was put to the Supervisory Board in March 1991. Early plans to contract the manufacture out to a single company — British Aerospace, CASA and Marshall's of Cambridge had been asked to tender — were put aside in favour of keeping the project in-house, and the go-ahead was given in August 1991. A new company, Special Aircraft Transport International Company (SATIC), was set up as a joint-venture between Aerospatiale and Deutsche Aerospace Airbus (now DaimlerChrysler), to manage the programme. Parts manufacture was shared out among the Airbus member countries, with final assembly contracted to Aerospatiale subsidiary Sogerma-Socea at its facility in Toulouse.

The Airbus A300-600ST Super Transporter, later given the more appropriate and evocative name of Beluga, shares an impressive commonality with the A300-600R, yet the giant whale-like aircraft bears little physical resemblance and is virtually a new aircraft, which differs, not only in size, but also in technical complexity. Construction of the Beluga begins on the normal A300-600R assembly line, before being completed by Sogerma-Socea. Some 80 percent of the airframe structure is identical, and includes the wings, General Electric CF6-80C2A8 turbofan engines, landing gear, most of the lower fuselage, and the pressurised flight deck, while system commonality is even higher at 90 percent.

The most notable structural changes are a new cavernous upper fuselage joined to the lower section at the normal wing/fuselage line, and a re-positioned flight deck below the main deck floor, to permit straight in loading through the largest single-piece, upward-opening freight door ever to be fitted to an aircraft. A side-opening door, as on the Guppy, was rejected early in the design development, as this would have created problems with the wiring. The new nose section is equipped with a galley, toilet and two additional passenger seats. The sheer size and weight of the Beluga required the stronger and larger tail assembly from the Airbus A340, with auxiliary fins fitted to the tips of the horizontal stabilisers to provide adequate lateral stability in cross winds up to 30kts (56km/h).

The result is a unique aircraft with the largest cross-section and volume of any cargo aircraft flying, at 24ft 3½in (7.4m) and 50,000 cu.ft (1,400m³) respectively. It also has two lower deck cargo compartments, one for bulk and one for containerised freight. In comparison with the Guppy, the Beluga carries 50 percent more volumetric load and is capable of transporting the largest single Airbus item, the front fuselage of the A330/340, measuring 18ft 4½in (5.6m) in diameter and 82ft (25m) in length. It has a payload of 47 tonnes, more than double that of the Guppy, sufficient to carry heaviest section, a complete wing

for the A340 model, which weighs in at 42 tonnes. The turbofan-powered aircraft is also considerably faster, and it is this speed advantage, together with a halving of the loading cycle through the use of special equipment installed at the manufacturing sites at Hamburg, Bremen, Chester and St Nazaire, that has enabled a reduction in the transport time of a complete Airbus from 45 hours to 19 hours. In other words, one Beluga represents the equivalent of 2.3 Guppies.

The first aircraft was assembled in 16 months, and on the morning of 13 September 1994, the A300-600ST took off from Toulouse–Blagnac on its maiden flight, which lasted for four hr and 21min. During the flight, the aircraft explored most of its range of operating speeds, altitudes and configurations with the undercarriage extended and retracted, providing a vital first check of behaviour under different conditions. The four crew members — flight test director and pilot in command Gilbert Defer, pilot Lucien Bernard, and Jean-Pierre Flamant and Didier Ronceray were unanimous in their praise of the aircraft, saying that 'the A300-600ST Beluga flew as we expected. Despite its unusual appearance, it behaved as a good and docile aircraft.' SATIC president Udo Dräger added: 'The successful first flight has proven that this aircraft can fit the bill . . . we are firmly convinced that the A300-600ST will find a place not only at Airbus Industrie but also in the international air cargo market'.

The 400 hours flight test programme ended with DGAC certification on 29 September 1995 and re-delivery to Airbus Industrie on 25 October. Entry into service took place in January 1996, after an extensive route-proving and loading site adaptation programme. Three more were delivered in March 1996, May 1997 and June 1998, and these four aircraft are considered sufficient to cope with a 350 aircraft per year production run, flying between 40 and 90 hours a week. A fifth A300-600ST is under construction for delivery in the year 2000 and

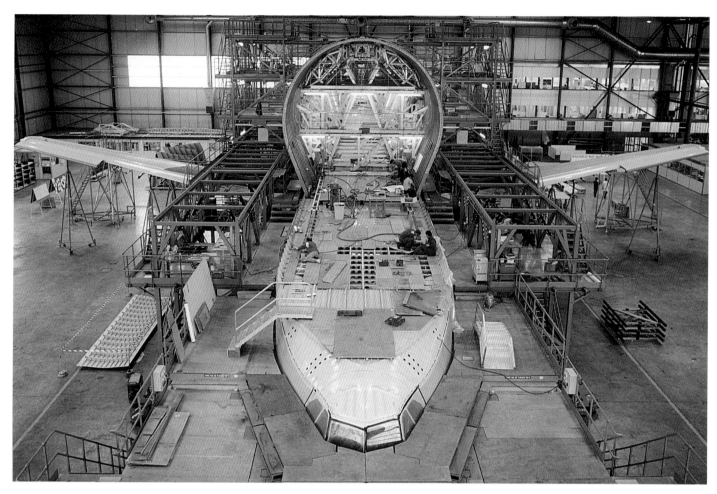

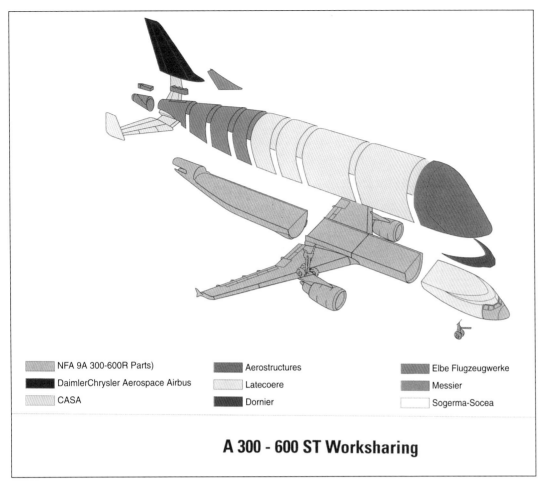

	NFA 9A 300-600R Parts)		Aerostructures		Elbe Flugzeugwerke
	DaimlerChrysler Aerospace Airbus		Latecoere		Messier
	CASA		Dornier		Sogerma-Socea

A 300 - 600 ST Worksharing

will be needed to cope with the increase in Airbus work, which is expected to triple over the next two years.

The present utilisation of the four Belugas on Airbus work still leaves around 2,000 hours per year, which are being sold to third parties on a charter basis through a new company, Airbus Transport International (ATI), set up specifically for this purpose. Charter flights began on 23 November 1996, when the Beluga transported a module for the International Space Station Alpha on behalf of Italian manufacturer Alenia Spazio. On 16 June 1997, the Beluga flew the world's largest-ever single piece of air cargo, a chemical tank built by Société Européenne de Chaudronnerie Industrielle (SECI), measuring 6.5m (21.3ft) in diameter and 17.6m (57.7ft) in length, and weighing 45 tonnes (44.2) including its supporting jig. A total of 18 such tanks will be transported between Clermont-Ferrand and Le Havre by June 1999.

The last of the four Guppies left Toulouse on 22 October 1997 on delivery to the National Aeronautics and Space Administration (NASA) at Houston, Texas, where it remains in regular use. The other three have been preserved, No.1 at Bruntingthorpe, UK, No.2 at Toulouse, and No.3 at Hamburg in Germany.

As for the passenger A300-600 and A300-600R, sales have slowed down. Those being completed on the final assembly lines are re-orders from existing customers. The September order from United Parcel Service (UPS) for 30 A300F4-600R freighters plus 30 options, will, however, keep the production line going for some years yet. At the beginning of 1999, the total A300 order book stood at 520 aircraft, of which 481 had been delivered. It is unlikely that a new member of the A300 family will be produced.

Few in the difficult early days, shaped so strongly by politics, the oil crisis and economic cycles, had given the European organisation and its aircraft much of a chance in the marketplace. But the A300 was just the beginning.

AEROSPATIALE	France	VU and Racks
		Floor Re-inforcement
AEROSTRUCTURES	Great Britain	Main Cargo Door
CASA	Spain	Cylindrical Upper Shells
DAIMLERCHRYSLER		Horizontal Tail Plane
AEROSPACE AIRBUS*	Germany	Cargo Loading System Installation
		Fin Re-inforcement
		Floor Re-inforcement
DORNIER*	Germany	Hydraulic System
		Structural Front Fairing
ELBE FLUGZEUGWERKE**	Germany	Conical Upper Shells
		Section 19.1
		Dorsal Fin
FOKKER	Netherlands	Water Drainage
		Insulation/Linings
		Miscellaneous Systems
		Floor Panels
LATECOERE	France	Nose Section
LHOTELIER	France	Air Conditioning System
RATIER FIGEAC	France	Mechanical Flight Controls
SOGERMA-SOCEA***	France	Main Longitudinal Joint
		Final Assembly
		Flight Test Installation
		Operational Support
		Return to Standard

* subsidiary of DaimlerChrysler Aerospace AG
** subsidiary of DaimlerChrysler Aerospace Airbus GmbH
*** subsidiary of Aerospatiale

5 AIRLINE OPERATORS

AEROCONDOR

Colombia's then second-largest international airline signed a letter of intent with Airbus on 15 July 1977, formalised into a firm order for an A300B4-102, plus one option, on 9 December. This aircraft was delivered the following day and entered service on the Barranquilla–Miami route on 15 December 1977. Fitted out for 245 passengers, the Airbus continued to serve the Miami route from Barranquilla, Bogotà, Cali and Medellin until repossessed by Airbus on 18 April 1979 and flown back to Toulouse on 5 May. Financial problems forced the airline out of business in May 1980. The A300 went on to enjoy a more secure existence with Air Inter.

AIR AFRIQUE

This multi-national West African carrier originally placed an order for one A300B4-203 and two A310s on 31 March 1979, but amended the order on 27 February 1981 by converting the A310s into A300s. The first was delivered to Abidjan, Air Afrique's Ivory Coast base, on 25 May 1981 and entered service on the routes to Paris on 2 June. The arrival of the second aircraft on 12 July 1983 and the third on 13 September 1984 enabled the replacement of older types on intercontinental routes from Abidjan and Dakar to Rome, as well as on West African links. In spring 1995, Air Afrique acquired two A300B4-605R on lease from ILFC.

AIR ALGÉRIE

The national flag-carrier of Algeria had a very brief encounter with the A300 as early as 25 November 1974, when it leased Trans European Airways' A300B1 for a series of Hadj flights. The aircraft was returned on 19 January 1975, and it was not until 1981 that a more permanent association was formed, albeit still on a lease arrangement. Two A300B4-2Cs were leased from Lufthansa, on 16 March 1981 and 31 March 1983, and used primarily on the busy Algiers–Paris sector. Both air-

RIGHT: Aerocondor A300B4-201.

BELOW RIGHT: Air Algérie A300B1.

BELOW: Air Afrique A300B4-203.

craft were delivered back to the German airline at the beginning of January 1985.

AIR FRANCE

To Air France goes the distinction of placing the first Airbus order. The French flag-carrier was involved in and has supported the programme from the very beginning, playing a major part in the final specification. From the time of the launch order, signed on 9 November 1971 for six A300B2-1Cs and 10 options, Air France steadily increased its fleet to eventually number 24 aircraft. The first was delivered on 10 May 1974, inaugurating the world's premier Airbus service on 23 May, with a flight on the busy Paris–London route. The delivery of seven A300B2-1Cs was followed from 16 November 1975 by 17 A300B4-203s, the last of which arrived in Paris on 15 April 1982. They served a mix of long-haul and medium-haul routes within Europe and to the Middle East and North Africa, equipped with either first and economy classes for 236 passengers, or a business and economy mix for 292 passengers. A300s were occasionally operated by its charter subsidiary Air Charter. Several of the early models were transferred to domestic carrier Air Inter, but these and other A300s came back to Air France on 1 April 1997, following Air Inter's full integration. However, with Air France ceasing A300 operations in 1997, these were sold on immediately, mainly for conversion to freighters.

AIR-INDIA

A sales tour of India in October 1973, during which the A300 made five demonstration flights at Bombay and Delhi, failed to attract an order from the flag-carrier, although domestic operator Indian Airlines ordered the type in April 1975. It was the A300B2 of Indian Airlines which gave Air-India its first experience with the Airbus, when it wet-leased an aircraft for its Abu Dhabi, Dubai and Muscat services between June 1977 and early 1979. When the airline's order for the TriStar was overtaken by Lockheed's decision to stop production, it urgently required a replacement and Airbus could offer virtually immediate delivery of A300 B4-203s, originally destined for Laker Airways. Air-India placed an order for three aircraft on 25 May 1982 and the first was handed over just two months later. The remaining two joined the fleet before the end of the year, and the Airbuses continue to fly some schedules to the Middle East and South East Asia.

AIR INTER

With France, along with Germany, the driving forces behind Airbus, internal airline Air Inter also ordered the A300. The first contract was signed on 24 December 1975, for three A300B2-1Cs, the type entering service on 8 November 1976 on routes linking Paris with southern France, serving Marseille, Toulouse, Nice, Bordeaux and Lyon. The firm order book increased to seven A300B2-1C models, but over the years, with more transferred from Air France, and others, including A300B4-2Cs, acquired from Lufthansa and Olympic Airways, a total of 22 A300s saw service with the French airline. 18 planes were still owned at the time of the integration into Air France on 1 April 1997, although the majority were either stored or leased out.

AIR JAMAICA

Jamaica's flag-carrier became the first Airbus customer in the Caribbean, when it signed a firm order for two A300B4-203 models on 9 January 1982 for delivery by October. They were to be used principally on the high density routes from Kingston to New York and Miami, as well as serving Toronto from both Kingston and Montego Bay. However, the order was never completed, with financial difficulties forcing the airline to opt instead for second-hand ex-Laker Airways aircraft, two of which were delivered to Jamaica on 20 February 1983. A third was optioned for 1985 but never taken up. Fitted out with 12

first class and 265 economy class seats, the A300B4-203 entered service with a Kingston–Toronto–Kingston roundtrip on 27 February and also served New York, Los Angeles and Miami. Two ex Brazilian-registered A300B4-203 were added in summer 1990, but all four left the fleet by October 1996 after the airline's re-organisation and part-privatisation. Air Jamaica also operated a fifth aircraft on lease from Airbus Industrie between 23 November 1992 and 3 October 1994.

AIR LIBERTÉ

This French charter and scheduled operator acquired two new A300-622Rs from International Lease Finance Corporation (ILFC) in spring 1990 to inaugurate long-haul services to Bangkok, Montreal, New York and Saint-Denis de la Réunion. Other routes were subsequently added to Saint Martin, Fort-de-France and Pointe-à-Pitre in the French West Indies, and one of the aircraft was leased to Garuda between 10 November 1990 and 1 April 1992. The A300s carried 345 passengers in a single cabin high-density layout. Both were returned to their lessor in October 1996.

AIR SEYCHELLES

An A300B4-200 was leased from Air France to replace a wet-leased British Caledonian DC-10-30 and was the first to carry full Air Seychelles colours. It entered service on weekend flights to Europe on 2 October 1985, linking Mahé with London, Frankfurt, Zürich, Rome and Jeddah. The Airbus continued operating the route until its return to Air France on 30 November 1987. It was again leased for a period in 1989.

AIR SIAM

Thailand's colourful, if short-lived, competitor to the national carrier became the world's second Airbus operator, when a leased A300B2-1C replaced a Boeing 747 on the Bangkok-Tokyo route on 21 October 1974. This aircraft was to be a stop-gap measure until the planned delivery of an A300B4 in summer 1975, which was to be followed by a second a year later. But the airline's ambitious expansion plans came to a stuttering halt through severe financial difficulties and neither aircraft was taken up. The leased A300B2 was returned to the

ABOVE: Air Jamaica A300B4-203.

BELOW: Air Liberté A300B4-622R.

BELOW RIGHT: Air Seychelles A300B4-203.

ABOVE RIGHT: Alitalia A300B4-203.

BOTTOM RGHT: Air Siam A300B2-1C.

manufacturer on 13 October 1975. Air Siam struggled on for another year, before being forced to cease operations on 1 January 1977.

ALITALIA

The Italian flag-carrier, a member of the Atlas consortium, had been targeted by Airbus from the very beginning as a likely customer, but it was not until 28 November 1978 a firm contract was signed (announced on 31 October) for eight A300B4-203s, plus three options. Four were delivered in 1981, starting on 28 April, three in 1981, and the last on 23 February 1982. The Airbuses entered service on Alitalia's routes from Rome to London, Paris, Cairo, Jeddah, Khartoum and Tel Aviv, and were also scheduled on the high-density Rome-Milan sector. All eight aircraft were configured for 18 first class and 251 economy class passengers. The fleet was enlarged to total 14 in autumn 1988, when Alitalia bought six ex-Eastern Air Lines aircraft, including two A300B2-202 and four A300B4-103 variants. Eight have since been sold.

AMERICAN AIRLINES

American became the launch customer for the extended-range A300-600R on 2 March 1987, with a huge order for 25 aircraft plus options, which brought the eventual fleet to 35. The first of the CF6-80C2A5-powered aircraft went into service with American in May 1988 and the airline took delivery of the 35th and last Airbus on 19 February 1993. Originally scheduled to various Caribbean destinations from the US mainland, the A300 is still used extensively into the Caribbean from its New York and Miami gateways, as well as on a few trans-Atlantic routes such as London Heathrow, London Gatwick and Paris Orly to Boston. It also flies London Heathrow–New York.

CHINA AIRLINES

In 1978, Airbus celebrated another customer in the Far East, when China Airlines placed an order for four A300B4-220s, plus an equal number of options on 1 September 1978. The order was subject to government approval, but this was not

forthcoming and the contract was rescinded the following month. However, on 30 July 1980, the deal was revived, this time for five aircraft, including one option, since taken up. Deliveries began on 22 June 1982 and the first A300 service was flown on 1 August 1982 on a Taipei–Manila–Taipei roundtrip. Between 25 September 1989 and 10 December 1992, China Airlines bought six new A300B4-622Rs, and added a further eight since December 1996, four of these leased from ILFC. Two were lost in tragic accidents in 1994 and 1998. The 18-strong Airbus fleet is now scheduled on the airline's South East and North East Asian routes.

CHINA EASTERN AIRLINES

As soon as the Civil Aviation Administration of China (CAAC) had decided to re-organise its civil aviation industry, it placed the first order with Airbus for its former Shanghai branch which, on 25 June 1988, became China Eastern Airlines. The new airline was authorised to serve the US and Japan markets and took delivery of 10 A300B4-605Rs between 24 November 1989 and 12 July 1995. China Eastern put the big Airbus onto its domestic trunk routes and regional services, now linking Shanghai to such destinations as Beijing, Guangzhou, Hong Kong, Shenzhen and Xiamen in China, and to Bangkok, Seoul, Singapore and Tokyo.

CHINA NORTHERN AIRLINES

This Shenyang-based airline was also the result of the re-organisation, but acquired its Airbus fleet much later, as it developed its regional network. The CAAC announced an order for six each for China Northern and China Northwest in April 1993, but the numbers were later reallocated. The first eight A300B4-622Rs arrived in Shenyang on 19 May 1993 and deliveries were completed within two years. Three are leased out to Korean Air, two of them delivered directly to that airline. The A300s fly high-density domestic and regional routes, fitted out, like the China Eastern aircraft, with 24 first class and 250 economy class passengers.

CHINA NORTHWEST AIRLINES

The third Chinese airline to receive the A300, China Northwest Airlines accepted the first two of five A300B4-605Rs on 6 and 14 October 1992. Both were ex-Compass Airlines aircraft leased from Polaris Leasing. Its three new aircraft arrived in China on 3 and 31 August 1994 and 23 November 1995. The Xi'an-based airline utilises its A300s on some of its mainline services to such destinations as Beijing, Hong Kong, Chengdu, Kunming, Shenzhen and Shanghai.

COMPASS AIRLINES

Deregulation in Australia gave birth to ambitious Compass Airlines, which secured early delivery positions for five A300-600Rs against a deposit lodged with leasing company Polaris in autumn 1989. In the interim, two A300-605Rs were leased from Monarch Airlines in the UK, both arriving in Brisbane in November 1990. Services began on 1 December with over 100

TOP: American Airlines A300B4-605R.

ABOVE LEFT: China Airlines A300B4-622R.

ABOVE: China Eastern Airlines A300B4-605R.

ABOVE RIGHT: China Northern Airlines A300B4-622R.

RIGHT: China Northwest Airlines A300B4-605R.

weekly flights between Brisbane, Sydney, Melbourne and Perth. But the aircraft proved too large for the market, with low load factors and an ensuing price war with the incumbent airlines leaving Compass floundering. Although the first two of the five new A300-605Rs were delivered in a 288-seat layout in April and August 1991, the writing was already on the wall and Compass suspended operations on 20 December that year. All four aircraft returned to their lessors the following February.

CONTINENTAL AIRLINES

The waning fortunes of competitor Eastern Air Lines provided the opportunity for Houston-based Continental Airlines to take some of that airline's A300s off its hands, as Eastern cut back its operations. It first acquired six A300B4-203s between 6 May 1986 and 8 July 1987, then added two ex-Singapore Airlines models of the same type in June and July 1986, before taking another 10 Eastern machines in 1990 and 1991 after that company had ceased operations. In the meantime, Continental had also placed an order with Airbus Industrie for three new A300B4-203s, which were delivered on 24 and 25 April, and

6 May 1986, bringing the fleet to 21 aircraft. The first Airbus services were flown in May that year out of Houston and Denver, replacing McDonnell Douglas DC-10-10s. After spending several periods in the shadow of Chapter 11 bankruptcy, the airline emerged a much stronger and leaner operation and continued to fly the Airbuses until disposing the fleet to leasing companies in 1995 and 1996.

CRUZEIRO DO SUL

Varig associate Cruzeiro do Sul signed a contract with Airbus on 27 April 1979 for two A300B4-203s and two options. The latter were converted to firm orders on 19 July 1979 and 30 September 1980. The first two arrived in Rio de Janeiro in June 1980 and entered service on 1 July linking Rio, Sao Paulo and Manaus in Brazil, and also serving international routes to Asunçion, Buenos Aires, Caracas and Miami. The second batch of two aircraft was re-assigned to Varig in October 1980, and Cruzeiro's machines were re-deployed mostly on domestic services, including a Porto Alegre–Fortaleza route, before leaving the fleet to go to Japan Air System and Air Jamaica on 15 March 1989 and 10 May 1990 respectively.

EASTERN AIR LINES

Such was Airbus Industrie's determination to break into the US market, especially after the snub by Western which had been expected to order the A300, that the manufacturer leased four aircraft to struggling Eastern Air Lines for a six-month period at no charge. Eastern was desperately short of capacity, but could not then afford to pay for new aircraft and the agreement hammered out with Airbus was a win-win situation for the US major. The first A300B4-103 was delivered on 24 August 1977 and, after several route proving, training and publicity flights, the Airbus was officially introduced with the winter schedule on 13 December, flying from New York La Guardia and Newark airports to popular winter destinations in Florida. Airbus Industrie's generosity paid off, as on 26 June 1978, Eastern signed for 23 A300B4-103s and took out options on a further nine. With further orders and conversions of options, the fleet had grown to 35 aircraft upon final delivery on 19 December 1983. This included 18 A300B4-103s, 15 A300B4-203s and two A300B2-202s. The A300s later uniformly seated 24 first class and 228 economy class passengers, with a few used in a 270 high-density layout on the New York-Boston shuttle. But the airline's fortunes failed to improve,

ABOVE: Compass Airlines A300B4-605R.

INSET LEFT: Continental Airlines A300B4-103.

INSET RIGHT: Eastern Air Lines A300B4-203.

leading first to Chapter 11 bankruptcy protection in March 1989, before operations were finally suspended on 18 January 1991.

EGYPTAIR

Since leasing a pair of A300s from Trans European Airways and Hapag-Lloyd from spring 1977, the Egyptian flag-carrier has become a prolific operator of Airbus types. It placed its own order on 29 April 1979 for three A300B4-203s, plus four options, two of which were converted to a firm contract on 16 October 1980, and the other two on 23 September 1981. An order for an eighth was signed later that year, such was the need for additional capacity to maintain its growing network of services. Deliveries of the eight aircraft were spread over three years, with the first arriving in Cairo on 19 September 1980. One A300B4-203 was leased from South African Airways between 27 August 1989 and 4 December 1990, during which

delivered to Dubai on 31 August 1993, but Emirates acquired a sixth, leased from ILFC on 6 March 1995. Emirates utilises its A300s on routes to Amman, Athens, Baku, Bangalore, Beirut, Cairo, Colombo, Damascus, Dhaka, Frankfurt, Istanbul, Karachi, Larnaca, Male, Mumbai, Muscat, Nice, Riyadh and Rome. Depending on which route is flown, the A300s are fitted out typically for 18 first, 35 business and 163 economy class passengers, or 21 business and 235 economy class passengers.

FEDERAL EXPRESS
The world's largest express freight transportation company placed a large order for 25 A300F4-600R freighters on 3 July 1991.

time EgyptAir started taking delivery of nine advanced A300-622Rs. Some of the earlier models have since been disposed off, but two remain in service following conversion to cargo. The A300-622Rs are fitted out for 253 passengers in a three-class layout.

EMIRATES
Dubai's newly-established flag-carrier used a wet-leased Pakistan International A300B4-103 from 1 November 1985 until 1 December 1988, followed by a Kuwait Airways A300C4-620, before taking delivery of the first of five new advanced A300B4-605Rs on 16 May 1989. The last of the five aircraft ordered from Airbus was

This provided a huge boost for Airbus which, until then, had only sold a handful of A300 cargo aircraft. Subsequent take-up of options brought orders to 36 aircraft. The first was delivered to FedEx at its Memphis 'superhub' on 27 April 1994 and just a few awaited delivery at the beginning of 1999. FedEx operates its Airbus-dominated fleet out of Memphis, Newark, Fort Worth, Indianapolis, Oakland, Chicago, Anchorage and Los Angeles, as well as international facilities in Europe, in Tokyo and Subic Bay.

TOP: EgyptAir A300B4-622R.

ABOVE: Emirates A300B4-605R.

LEFT: Federal Express A300F4-605R.

ABOVE RIGHT: Finnair A300B4-203.

RIGHT: Karair A300B4-203.

FINNAIR

The Finnish flag-carrier ordered two A300B4-203s for its associate Karair, which were delivered on 12 December 1986 and 13 March 1987. The first service was flown on 17 December 1986 with flights from Helsinki to Arrecife, and from Las Palmas de Gran Canaria back to Helsinki. Fitted out for 308 passengers, the two A300s achieved the highest utilisation of any Airbus then flying, logging 12 hours a day on charter flights from Finland to the Mediterranean resorts and the Canary Islands. Both aircraft were repainted in Finnair colours following the take over of Karair's charter operation on 1 April 1990, and remained in service until the last flight from Helsinki to London Stansted on 26 April 1998. Both were sold to Air Scandic and operate in the UK.

EUROPEAN EXPERIMENTS

The Dutch charter operator Transavia leased the ex-Air Siam A300B2 from Airbus as early as 6 May 1976, pending delivery of its own aircraft ordered the day before. However, in October that year the order was cancelled in favour of the smaller Boeing 737 and the aircraft returned to the manufacturer on 18 January 1977. Another short-lived Airbus operator was Ratioflug, which acquired an ex-Air France A300B2-1C on 9 May 1997, but ceased operations seven months later. The Frankfurt-based charter company operated the A300 in a 323-seat tourist layout on holiday charters flights to the Mediterranean resorts from Düsseldorf.

Two UK holiday airline operators of A300s are also no more. London Gatwick-based Dan-Air Services added extra capacity in the 1980s, with the lease of two A300B4-103s from German company Hapag-Lloyd in April and December 1986, followed by a third in March 1988. All had been returned by December 1989. Orion Airways acquired two A300B4-203s from Lufthansa in spring 1987 for its Mediterranean charters, mostly out of East Midlands Airport, but both came nominally into Britannia Airways ownership when Orion was integrated in November 1988. Britannia never operated the type in its own colours prior to their sale to Iberia on 28 April 1989.

Two A300B4-203s flown by UK charter company Peach Air in 1998 were owned by Air Scandic and operated on its behalf by Luton-based Air Foyle. The first service was flown on 1 May 1998 between Manchester and Tenerife. Other holiday destinations served regularly are Arrecife, Dalaman, Larnaca, Malaga and Paphos. The Air Scandic aircraft are also used in its own colours. The fleet of European Air Charter (EAC) now includes two A300B2-1Cs from Air France, and an ex-EgyptAir A300B4-203F freighter, all bought at the beginning of 1998. These are mainly wet-leased to other operators, as are the six A300s (two A300B4-103s, three A300B4-203s and one A300B4-203F) acquired from various sources by Irish operator TransAer (formerly Translift Airways) between 1995 and 1997, and based at Dublin, Manchester and London Gatwick.

Luxair operated a single ex-Singapore Airlines A300B4-203 on behalf of Luxavia from 19 December 1984 until its sale to South African Airways on 20 December 1987.

TOP RIGHT: Orion Airways A300B4-203.

BELOW RIGHT: Dan-Air A300B4-103. *Mike Irish*

BELOW: Air Scandic A300B4-203. *Terry Shone*

FREIGHTER FOCUS

Although Airbus had delivered a handful of all-cargo models to various airlines, the big freighter breakthrough came on 3 July 1991 when US carrier Federal Express (FedEx), the world's biggest express cargo airline, signed a contract for 25 dedicated A300F4-606Rs, since increased to 36 aircraft. The first was handed over on 2 December 1993 and deliveries will be completed during 1999. Airbus scored a further notable success during the 1998 Farnborough Air Show when another prominent US package delivery company, United Parcel Service (UPS), placed its first order with Airbus for 30 A300F4-600s and a similar number of options, for delivery from mid-2000.

Meanwhile, older A300B2s and A300B4s were becoming available and a growing cargo conversion programme has been initiated by British Aerospace Aviation Services in the UK, Dasa in Germany and Sogerma in France. The Bournemouth, UK-based cargo airline Channel Express became the launch customer for the programme, acquiring its first A300B4-100 for conversion in 1996. The first A300B4-103F, converted by British Aerospace at Filton, entered service in July 1997, and Channel Express subsequently introduced two A300B4-203Fs, converted by Dasa in Dresden. The A300s, referred to as the Eurofreighter, operate

throughout Europe and the Middle East. Another UK cargo operator, HeavyLift Cargo Airlines, is leasing three A300B4-203Fs, first delivered on 10 November 1997.

Other cargo aircraft converted by British Aerospace have gone to Turkish carrier MNG Cargo Airlines, to Jetlink International of the Netherlands, to ACS Cargo and JHM Cargo Express, both in Costa Rica. The first of seven for JHM was handed over to the customer during the Farnborough Air Show in September 1998.

DHL International, the European arm of the US express parcels carrier has steadily built up its fleet of converted A300B4-200F freighters to nine, which are operated by Hunting Cargo Airlines (Ireland), EAT in Belgium, and Switzerland's Farnair Europe on its behalf. Farnair Europe acquired three Dasa conversions in spring 1998.

ABOVE LEFT: DHL converted A300B4-200F freighter.

LEFT: TNT A300B4-200F.

BELOW: JHM Cargo Express A300B4-200F.

ABOVE: Garuda Indonesia A300B4-622R. *Günter Endres*

RIGHT: Germanair A300B4-2C.

GARUDA INDONESIA

The Indonesian flag-carrier, then known as Garuda Indonesian Airways, placed an order for six A300B4-220s, with a similar number of options, on 24 February 1979. That order also launched the forward facing crew cockpit (FFCC) and Garuda's first aircraft was also the first wide-body aircraft to fly with a two-man crew when it took to the skies on 6 October 1981. The six aircraft were delivered to Jakarta within a two-month period starting on 11 January 1982, and entered service on Garuda's high-density domestic routes, such as Jakarta–Medan and Jakarta–Denpasar Bali. On 18 March 1981, Garuda had converted three of its options into firm orders, which joined the fleet in November 1982. Between 22 August 1991 and 2 March 1993, the airline leased a total of 10 A300-622Rs from various sources, but all have been returned. After losing one aircraft in a highly-publicised accident in September 1997, the operational A300B4-220 fleet now numbers eight.

GERMANAIR

Charter company Germanair — until 1968 known as Südwestflug — was an early customer for the A300, ordering one A300B4-103 plus one option on 11 February 1975. It received its aircraft just three months later, and flew its first service 1 June 1975, becoming the first German Airbus operator, ahead even of Lufthansa. A second joined on 1 April 1976 and the two aircraft, fitted out with 314 seats, became popular on the airline's holiday routes to the Mediterranean and Canary Islands resorts. After joining operations with Bavaria Fluggesellschaft under the same ownership on 1 January 1977, the new Bavaria–Germanair doubled the fleet, taking delivery on 2 May 1977 and 20 February 1978, although one was leased to Egyptair. All four became part of the Hapag-Lloyd fleet, following a take-over in 1979.

HAPAG-LLOYD

Germany's second-largest holiday charter airline, based at Hannover, began Airbus operations in 1979, when it inherited four A300B4-103s from the take-over of Bavaria-Germanair. All were converted to A300B4-203 standard. Hapag-Lloyd had itself ordered the type on 9 May 1977, signing up for one A300B4-203 plus one option, and on 31 July 1978 became the launch customer for the convertible A300C4-203. The latter, equipped in a 315-seat one class cabin in common with the other models, was operated on passenger charters following its delivery on 31 January 1980, but re-configured to cargo mode in January 1982. This aircraft logged two notable records that year, becoming the first wide-body twin on 23 January to be used on pure cargo work, and the first A300 to make a commercial flight from Europe to Australia on 14 June. After two passenger aircraft returned from long-term lease to EgyptAir in

1982, the airline found itself with surplus capacity. One of the seven-strong fleet was sold to Pakistan International, while another went to America on long-term lease to Capitol Air. It was the beginning of a gradual process to replace the A300s with the smaller A310. One more went to PIA in April 1986, two ex-Singapore Airlines aircraft which had been acquired on 28 June 1985 were sold to Toa Domestic in March 1988, three were disposed of to Guinness Peat during 1989, and the last to Air Inter on 9 January 1990.

IBERIA

The Spanish flag-carrier was the second Airbus customer when it signed an order for four A300B4-103s, plus one option, on 14 January 1972, in the process becoming the launch customer for the extended-range B4 model. This early order was not unconnected, with Spain (through CASA) joining the Airbus consortium around the same time, and it came as a blow to the manufacturer when the order was later cancelled. Nearly seven years had elapsed when Iberia came back to Airbus, contracting

ABOVE: Hapag Lloyd A300B4-103. *David Hill*

BELOW: Iberia A300B4-120.

ABOVE RIGHT: Indian Airlines A300B2-101.

BELOW RIGHT: Iran Air A300B4-605R.

to buy four Pratt & Whitney-powered A300B4-120s on 28 December 1978. An option on four more was also taken out. The order was amended on 3 October 1979 through the conversion of the four options, an order for an additional aircraft, and five new options, although subsequent cancellations

brought the acquisition back to six. Iberia took delivery of its aircraft over a one year period between 27 February 1981 and 26 February 1982. The first service with Iberia was operated on the delivery flight from Toulouse to Madrid (Getafe). Two ex-Orion Airways A300B4-203s were bought on 28 April 1989 and all eight aircraft remain in service and regularly link Spain with destinations in Europe, the Middle East and North Africa.

INDIAN AIRLINES
Government-owned Indian Airlines became the sixth A300 customer, signing for three A300B2-1Cs, plus three options, on 24 April 1975. The arrival of the wide-body A300 helped the

airline to eliminate some of its chronic capacity problems and on 1 December 1976, the airline put the Airbus onto its trunk routes linking Bombay, Delhi, Calcutta, Madras and Bangalore. On 2 April 1978, Indian Airlines took up two options and took out three more, later taken up, which brought the fleet to eight, with the final delivery on 7 August 1980. Two new A300B4-203s were delivered in May 1982, followed by two ex-Lufthansa A300B2-1Cs, bought from Airbus on 22 August 1982. Indian Airlines' 10-strong fleet (two were lost in accidents) is scheduled primarily on major domestic routes, serving Bangalore, Calcutta, Chennai (Madras), Delhi, Goa and Mumbai (Bombay), fitted out for 248 passengers in two-classes.

IRAN AIR

The first Airbus order from the Middle East came on 4 March 1978, when Iran Air contracted to buy six A300B2-200 models and took out three options, although these were not taken up. Pending delivery of these aircraft, Iran Air also signed a lease deal with Airbus for two more, which arrived in Tehran on 7 and 16 March 1978 and remained in service until 1 January 1979. Its own aircraft were delivered between 17 March 1980 and 31 January 1983 and operated in single-class configuration on the main domestic routes from Tehran to Abadan, Mashad, Shiraz and Zahedan, but were soon enmeshed in the Iran-Iraq war. One Airbus was impounded at Iraq after a hijack on

of Asian customers for the Airbus, when it placed a contract on 21 May 1979 for six A300B2K-3Cs, specially-adapted with Krüger flaps for improved performance. On 21 September that year, TDA increased its commitment to nine and took out three options. The nine aircraft were delivered between 2 October 1980 and 16 June 1983, first entering service on the Tokyo–Kagoshima and Fukuoka–

28 August 1984 and not returned until 15 September 1990, while a second was shot down (inadvertently) by the US Navy over the Gulf after take-off from Bandar Abbas on 3 July 1988. On 27 December 1994, the fleet was boosted by two new A300-600Rs, fitted out for 22 business class and 239 economy class passengers. The remaining seven aircraft can be seen regularly at Ahwaz, Bandar Abbas, Bushehr, Dubai, Esfahan, Frankfurt, Hamburg, Kerman, Mashad, Shiraz, Tabriz and Zahedan.

JAPAN AIR SYSTEM

Japan's third-largest airline, until 1 April 1988 trading as Toa Domestic Airlines (TDA), joined the growing list

Kagoshima routes fitted out for 281 passengers in a high-density eight-abreast arrangement. The A300s served the airline well, leading to a decision on 29 July 1985 to add some second-hand A300B4-203s, and a total of eight (including an A300C4-203 not taken up by Libyan Arab Airlines) were delivered between 1987 and 1991. Another momentous decision was taken by Japan Air System on 30 March 1989 with an orders and options for 17 larger A300-622Rs, since increased to 19 firm orders. Deliveries started on 25 April 1991 and will be completed in 1999, at which time the A300 fleet will total 36 aircraft and will be the largest in airline service.

KOREAN AIR

South Korea's privately-owned national airline started off the order spree after certifi-

cation, by signing a contract for six A300B4-103s on 5 September 1974. Two of these were conditional, but were confirmed on 27 January 1975. Iberia's cancellation gave Korean Air Lines early delivery positions, with the first aircraft arriving in Seoul on 1 August 1975 and boosting the airline's growing regional network, serving routes to Japan, Taiwan and the Philippines. The first service was operated on 28 August 1975. Deliveries of all seven aircraft — a seventh had been ordered on 4 March 1978 — were completed on 9 August 1978. Others were added in subsequent years, including one more A300B4-103, one convertible A300C4-203, and a launch order on 11 October 1985 for two A300F4-203 all-freighters. The airline then went over to Pratt & Whitney-powered models, first acquiring some A300B4-620 and -622 models, before placing a massive launch order on 11 March 1988 for 16 A300-622Rs, which were delivered between 29 November 1988 and 26 May 1994. The airline is now one of the biggest A300 operators, with 29 in service on routes from Seoul.

KUWAIT AIRWAYS

The Gulf states were very much Boeing territory until a break-through order for Airbus was announced by Kuwait Airways on 27 June 1980. The order was signed on 6 December for six A310s, plus five options, but on 30 June the following year, two of these options were confirmed, with the remaining three changed to the A300C4-620 convertible, for which Kuwait Airways became the launch customer. It took delivery of the first aircraft on 30 May 1984. On 2 August 1990, all three were seized by Iraq and flown to Baghdad, where two were destroyed by Allied bombing at the height of the Gulf War on 15 February 1991. The other was returned to Kuwait in September 1990 and joined in 1993/94 by five General Electric-powered A300B4-605R models. The six aircraft now form part of an Airbus-dominated fleet, linking Kuwait to such destinations as Abu Dhabi, Beirut, Cairo, Damascus, Dhahran, Delhi, Frankfurt, Jeddah, London, Madrid, Muscat, Paris, Riyadh, Rome, Tehran and Trivandrum.

LAKER AIRWAYS

This holiday and low-fare trans-Atlantic carrier headed by colourful entrepreneur Freddie Laker (now Sir Freddie) placed an order for 10 A300B4-200s on 10 April 1979, but only three were delivered before the collapse of the airline in February 1982. The first aircraft, configured for 298 passengers, was introduced on the London Gatwick–Palma de Mallorca and Gatwick-Monastir flights on 10 January 1981. All three were on strength by 11 June, flying holiday charters from Gatwick and Manchester, but were placed into open storage at London Stansted when Laker Airways folded. Two found their way to Air Jamaica on 18 February 1983, and the third went to Pakistan International in June 1984.

LATIN LEASES

Venezuelan flag-carrier Venezolana Internacional de Aviacion SA (VIASA) leased A300B4-203s from GPA in summer 1987. The two aircraft, both ex-Lufthansa machines, flew into Caracas on 20 August and 21 September 1987 and stayed with the airline until December 1994, serving the routes to Curaçao, Houston, Miami, New York, Rio de Janeiro, Santo Domingo, San Juan and Toronto. They were configured in a comfortable two-class layout for 23 club and 209 economy passengers. VIASA has since ceased to operate. Another airline no longer in existence, Faucett of Peru, leased two A300B4-203s on 15 December 1994 and 11 April 1995, but almost immediately wet-leased both to APA International Air of Santo Domingo in the Dominican Republic. APA International Air used the aircraft on

its scheduled combination routes from Santo Domingo and Puerto Plata to Miami and New York, via San Juan, Puerto Rico. One of the A300s had earlier spent a short period with Antiguan-registered Caribjet. Faucett also used two more on its own services.

With the two major airlines, Aeromexico and Mexicana, having both opted for the DC-10 in the 1970s, the A300 did not appear in Mexico until 3 August 1989, when now defunct holiday charter airline Latur leased an A300B4-622 from Airbus Industrie to fly US leisure travellers to the Mexican resorts. A second, this time an extended-range A300B4-622R, was bought new and delivered on 13 December that same year. Both went to Garuda in early 1990 and the airline ceased flying on 18 December 1991.

The A300B4-622R found its way to another charter operator Aerocancun in Cancun, where it served from 16 December 1994 until 10 May 1996. A third Mexican operator is Taesa, which leased two A300B4-203s from ING Aviation Lease for its US flights on 1 July 1995 and 13 December 1995. One was returned on 12 May 1996, while the other remains in service, fitted out in a 320-seat charter layout.

Ever-present Dominicana, based at Santo Domingo in the Dominican Republic, operates entirely with leased aircraft as and when required. Among these were two A300B4-120s, leased for a short period from Danish charter airline Conair between December 1992 and 1 February 1993.

ABOVE LEFT: Varig PP-VND.

ABOVE: Latur A300B4-622.

LEFT: Ladeco A300B4-203.

LIBYAN ARAB AIRLINES

Colonel Ghaddafi's Libya was no friend of the Americans, and when the national airline placed an order for Airbus aircraft on 15 September 1981, which included four A300B4-203s, two A300C4-203s and four A310s, the US stepped in to halt the sale. US companies were banned from supplying equipment which could be used for military purposes, and as the aircraft had a high US content, which included the engines, none were released. Consideration was given to installing the Rolls-Royce RB211, but the cost of the redesign and certification was too prohibitive. The six A300s were eventually completed, with the four A300B4-203s leased to Pan American, and the A300C4-203s going to Korean Air Lines and Toa Domestic Airlines.

LUFTHANSA

In spite of Germany's large involvement in the Airbus consortium, the German flag-carrier was a reluctant early customer for the A300, having always believed that a smaller type would be more suitable. Nevertheless, and presumably under some pressure from its government owner, Lufthansa signed a contract for three A300B2, plus four options, on 7 May 1973, becoming the third airline to order the new type. The first A300B2-1C was handed over to the airline on 2 February 1976 in time for service entry on the Frankfurt–London and Frankfurt–Paris routes on 1 April. All three were delivered by 25 April, enabling the airline to put the type also on the Madrid run and onto main domestic trunk routes serving Düsseldorf, Hamburg, Hannover and Stuttgart. Three more A300B2-1Cs and five A300B4-2Cs followed, but all had left the fleet by 1988, at which time, Lufthansa was taking delivery of 11 new A300B4-603s. Two extended-range A300B4-605Rs joined during 1996. The earlier A300B4-203s were on occasions leased to charter subsidiary Condor Flugdienst.

MALAYSIA AIRLINES

The Malaysian flag-carrier, then trading as MAS (Malaysian Airline System) ordered three A300B4-200s on 30 August 1978, also taking an option on a fourth, later converted to a firm contract. The first aircraft was delivered on 30 October 1979, with the next two arriving at Kuala Lumpur on 28 December, and 18 January 1980. The fourth did not join the fleet until 24 July 1981. The Airbuses entered service on 16 November 1979, first to Jakarta, Hong Kong,

LEFT: Lufthansa A300B4-603.

Madras and Perth, then, on 1 January 1980, to Taipei, Tokyo and Seoul. They also worked hard on the busy Kuala Lumpur–Singapore route, as well as on high-density domestic services. Three were sold to a leasing company and went to Carnival Air Lines in the United States between December 1994 and July 1995, while the fourth has been on lease to Air Maldives since 10 November 1994.

MONARCH AIRLINES

UK charter company Monarch Airlines, based at London Luton Airport, placed an order for four A300B4-605Rs on 5 February 1989 and received the first two on 15 March and 17 April 1990, with the other two arriving on 26 April and 3 May the following year. Apart from the lease of two aircraft to Compass Airlines of Australia between November 1990 and January 1992, the type has been in continuous service with the Luton-based holiday airline and is the mainstay of the widebody fleet. The four aircraft achieve very high utilisation on Monarch's Mediterranean and US holiday routes, where they are fitted out for 361 passengers in what is the highest certificated capacity of the A300-600R.

OLYMPIC AIRWAYS

The choice of a wide-body jet for Olympic Airways' high-density European services fell on the Airbus A300B4-103, and the two aircraft of the initial order were handed over at Toulouse on 22 February 1979. Three additional A300B4s were subsequently ordered for delivery in March and April 1980, which enabled more European Airbus destinations to be added. A sixth joined the fleet in 1981. Two A300B2-1Cs were also leased from Airbus Industrie for around a year in 1985/86. A need for yet more capacity and a desire for further modernisation of the fleet was behind the order for two improved A300-605Rs. The first arrived in Athens on 5 June 1992, followed by the second on 4 October 1993. All eight remain in service.

RIGHT: Monarch Airlines A300B4-605R.

INSET LEFT: Malaysia Airlines A300B4-203. *Günter Endres*

INSET RIGHT: Olympic Airways A300B4-605R.

MEDITERRANEAN MIX

The high capacity and low operating costs of the A300 have attracted many holiday airlines, with the Mediterranean a particulary fertile hunting ground for the big Airbus. The A300 has featured regularly in the frequently-changing charter industry in Turkey. Sultan Air acquired two ex-Eastern Air Lines A300B4-203s on lease from Polaris on 13 December 1992 and 15 March 1993, but ceased operations in October. Akdeniz Airlines leased three ex-Continental A300B4-103s on 1 June 1995, but stopped flying the following October. A similar fate befell Holiday Airlines, which began A300 operations in April 1995 with an Air Inter A300B4-2C, and subsequently acquired two more from the same source. All three were returned on 1 October 1996, a few weeks before the airline went out of business. Birgenair had a Finnair A300B4-203 on short-term lease in September 1995, but became yet another to go to the wall.

But the A300 can still be found in service with more permanent Turkish carriers. Air Anatolia flies four A300s, including an A300B2-1C, delivered on 8 July 1997; an ex-Thai A300B4-103 delivered on 19 September 1996, an ex-Air Jamaica A300B4-203, delivered on 22 March 1997, and an A300B4-2C leased from Onur Air. Air Alfa has five A300s — two A300B4-103s and three A300B4-203s — taken on between May 1994 and May 1998. Another A300B4-203 was leased between 29 April 1996 and 10 February 1997. Two of these had earlier been in service with Belgian carrier European Airlines, which also leased another, before going out of business on 1 January 1996. Onur Air took over two A300B4-103s from defunct Akdeniz Airlines, and also acquired an A300B4-2C from Air France on 6 June 1997. Pegasus Airlines also leased an ex-Carnival/Pan Am aircraft for a short period.

Apollo Airlines, a short-lived charter company in neighbouring Greece, utilised three A300s on charter flights mainly from Scandinavia. All three A300B4-203s, one leased from ING Aviation and two from Airbus Industrie were delivered in 1995, starting on 3 March, but the airline suspended operations on 1 December 1996. Another airline carrying holidaymakers from Scandinavia to the Mediterranean resorts was Swedish-based Air Ops.

Although the company operated a fleet of Lockheed TriStars until its collapse on 30 April 1996, it also had three A300B4-203s on short-term leases between June and December 1995. ZAS Airline of Egypt leased three A300B4-120s from Conair of Scandinavia between spring 1991 and the end of 1992, and replaced them with two ex-Air France A300B4-203s in March and April 1993. However, both were repossessed on 1 November 1995 after the airline ceased operating. One of the two stayed in the country and went into service with AMC Aviation, where it carries up to 287 passengers on charters to the local historical and holiday sites.

ABOVE: Onur Air A300B4-103.

TOP LEFT: Sultan Air A300B4-203.

LEFT: Apollo Airlines A300B4-203.

NORTH AMERICAN NOTES

Florida-based Northeastern International Airways was the second US airline to operate the type, leasing two surplus Lufthansa A300B2-1Cs on 24 January and 4 February 1984. The airline commenced Airbus service on 9 February on passenger charters, serving the Florida resorts of Fort Lauderdale, Orlando, Tampa/St Petersburg and West Palm Beach from the New York and Boston area. A scheduled Miami–New York service was introduced on 2 May. But Northeastern was not in good financial health and, although leasing two more from the manufacturer on 29 May, all four were returned during October. The airline ceased operations the following January.

Long-established Capitol Air was another airline, which was to cease operations soon after taking a long lease on a Hapag-Lloyd A300B4 in spring 1984. It used its Airbus principally on its New York–Chicago–Los Angeles schedule but stopped flying on 26 October 1984. The same fate befell low-fare carrier Presidential Air, which acquired three ex-Continental Airlines A300B4-203s on lease from ING Aviation Lease on 1 September 1995 and launched daily charter flights on 2 October from Long Beach to Houston and Atlanta on behalf of Presidential Tours. The end came on 9 February the following year, when all three aircraft were taken back by the lessor. Two of these then joined the fleet of Carnival Air Lines, which had already taken delivery of six A300B4-203s in 1994 and 1995, with three ex-Malaysia Airlines, one ex-Eastern, and two formerly operated by the original Pan American World Airways and acquired from Airbus Leasing. The A300s were used on the high-density domestic scheduled and charter routes, as well as to Lima in Peru.

In July 1996, the new Pan Am (the trade name was purchased in 1993), announced that it would buy Carnival Air Lines, although that airline continued to fly under its own colours for another year. Pan Am itself acquired six A300B4-203s on lease, beginning operations in August 1996 from Miami, serving the Northeast, Midwest, Florida, the Bahamas and the Caribbean. The A300s were scheduled on the longer and busier routes, fitted for 24 first and 230 economy class passengers. But there was to be no renaissance and the once famous name failed to save the airline. All operations were suspended on 27 February 1998.

Prior to taking delayed delivery of its six A310-300s ordered on 25 March 1981, charter carrier Wardair Canada operated two A300B4-203s and one A300C4-203 on lease from South African Airways. All three were delivered during August 1986, with the last returned to the lessor on 3 May 1989, following the acquisition of the airline by the PWA Corporation. The A300s operated mainly on its Florida and Caribbean routes, fitted out for 30 first class and 220 economy class passengers.

ABOVE: Wardair A300B4-203. *Keith Sommer*

TOP LEFT: Northeastern International Airways A300B2-1C. *Maurice Bertrand*

LEFT: Carnival Air Lines A300B4-203.

PAKISTAN INTERNATIONAL AIRLINES

The Pakistan flag-carrier ordered four A300B4-203s on 16 July 1978 and took out options on six more, none of which were taken up, although PIA has since acquired six second-hand models from various sources. The four aircraft from the original order were all delivered during 1980, with the first arriving in Karachi on 3 March 1980. The Airbuses are equipped for 24 first class and 222 economy class passengers and are scheduled on routes from Karachi to such points as Abu Dhabi, Al Ain, Bahrain, Bangkok, Colombo, Delhi, Dhahran, Dhaka, Doha, Dubai, Islamabad, Kuwait, Mumbai (Bombay), Peshawar and Quetta.

PAN AMERICAN

The announcement on 13 September 1984 that Pan Am had signed a letter of intent for 12 A310-300s and 16 A320s came as a welcome boost to the European consortium. As an interim measure, Airbus Industrie offered the airline an early delivery of 12 A300B4-203s then in store, and an initial contract was signed with financial institutions for the first four on 21 December 1984. These were delivered on the same day and entered service on US and Caribbean routes. Pan Am took delivery of the remaining eight aircraft between March and May 1985, and acquired a 13th on lease from United Aviation Services on 21 January 1990. However, the famous airline was

already struggling to survive and finally succumbed to financial pressures on 4 December 1991.

ABOVE: Pan American A300B4-203s.

INSET: Philippine AirlinesA300B4-103.

PHILIPPINE AIRLINES

Under a new leadership determined to improve service and equipment, the Philippine flag-carrier ordered two Airbus A300B4-100s, plus two options, on 15 November 1978 for its South East Asia routes. Both were delivered at the end of 1979, with the first entering service on the Manila–Singapore service on 4 December. The options were soon taken up and another ordered, bringing the fleet to five, including two A300B4-100s and three -200s, when deliveries were completed on 29 March 1983. Since then, Philippine Airlines has leased several more from GECAS and Airbus Industrie, including four in spring 1995. The fleet totals 12 aircraft, which mainly served regional routes to Fukuoka, Hong Kong, Ho Chi Minh City, Jakarta, Osaka, Seoul, Singapore and Taipei, and the local trunk routes from Manila to Cebu and Davao, until the airline stopped flying in autumn 1998. Limited operations have recommenced.

QANTAS

The Australian flag-carrier inherited four A300B4-203s from Australian Airlines, following takeover of the domestic airline on 31 October 1993. Fitted out for 250 passengers, including a first class cabin with 30 seats, the Airbuses were scheduled on high-density routes linking the state capitals of Sydney, Brisbane, Hobart and Perth, with Launceston on Tasmania also served. All four were sold for cargo conversion in autumn 1998.

QATAR AIRWAYS

Now the flag-carrier of the State of Qatar, the airline was re-organised in spring 1997, introducing three ex-Garuda A300B4-622Rs leased from Ansett Worldwide Aviation Services. Qatar Airways took delivery of the three aircraft on 30 March, 28 April and 10 December 1997, and operates the type in a three class configuration for 231 passengers, claiming the widest seat pitch in the Middle East. The A300B4-622Rs are scheduled on its routes from Doha to Munich and London, and eastwards to Colombo, Dhaka, Karachi, Kathmandu, Mumbai (Bombay), Peshawar and Trivandrum.

SAUDI ARABIAN AIRLINES

The Saudi flag-carrier, then trading as Saudia, joined the ranks of Airbus operators in 1980 when it introduced two A300B2-103s on a one-year lease from Korean Air Lines in November 1980. The two aircraft were used on its Arabian Express services linking Jeddah, Riyadh and Dhahran, fitted out with 24 first class and 226 economy class seats. On 12 December 1980, Saudia signed for 11 of the larger, advanced A300B4-620s, powered by the Pratt & Whitney JT9D-7R4H1. The engine

ABOVE LEFT: Scanair A300B2-320. *W.F. Wilson*

BELOW LEFT: Qantas A300B4-203.

BELOW: Saudia A300B4-620.

choice was a disappointment for Rolls-Royce and Airbus, both of which had hoped that Saudia would choose the RB211 to provide commonality with other aircraft in the fleet. All 11 aircraft were delivered during 1984, the first service being flown on 14 May. The aircraft are configured for 26 first class and 232 economy class passengers and are used mainly on regional services and the high-density domestic trunk routes.

SCANDINAVIAN AIRLINES SYSTEM/SCANAIR

SAS announced in mid-1977 that it was evaluating the A300 for use on its own scheduled services and for charter subsidiary Scanair, which led to an order for two A300B2s, plus 10 options, on 30 December that year. Two options were taken up on 2 March 1979, but the rest were subsequently cancelled. To meet the airline's requirements, Airbus incorporated several modifications and fitted the P&W JT9D-59A, resulting in an aircraft somewhere between the B2 and B4 models. Designated A300B2-320 and bought only by SAS, the first aircraft was delivered on 15 January 1980 entering service on the Copenhagen–London and Copenhagen–Oslo routes on 18 February. With deliveries completed on 12 March 1981, two aircraft were operated on mainline scheduled services in a mixed 238-seat layout; the other two served on Scanair's charter routes in a 291-seat high-density arrangement. To improve performance, all four were returned to Deutsche Airbus in Bremen and converted to B4-120 standard, the main work being the installation of a centre fuel tank for greater range.

However, operating such a small fleet proved uneconomic and the aircraft were frequently leased out to other operators. It was during a short lease with Malaysian Airline System (MAS) that one crashed on approach to Kuala Lumpur in bad weather on 18 December 1983 and burned out, luckily without the loss of lives. The remaining aircraft were sold during 1987 to Conair of Scandinavia, which became Premiair on 1 January 1994 when merged with Scanair. All three remain in service.

99

SOUTHERN SIGHTS

The Indonesian scheduled and charter carrier Sempati Air leased three ex-Pan American A300B4-203s from Airbus Leasing, all three being delivered in 1993, starting on 23 April. A fourth joined the fleet on 20 June 1996. Sempati used the Airbuses on its high-density domestic services linking Jakarta, Balikpapan, Denpasar Bali, Medan and Surabaya, as well as on regional flights to Bangkok, Kaohsiung and Singapore, until becoming one of the first victims of the Asian economic crisis and ceasing operations in early 1998. Garuda associate Merpati Nusantara has been leasing a Kuwait Airways A300C4-620 since 1 April 1997.

In neighbouring Philippines, ambitious Grand International Airways, trading as GrandAir, quickly built up its Airbus fleet, taking delivery of an A300B4-2C on 17 March 1995, and following up with four A300B4-203s over the next two years. All were leased from various sources. The airline operated the

ABOVE: Grand International Airways A300B4-2C.

RIGHT: Sempati Air A300B4-203.

BELOW: Air Niugini A300B4-203 .

Airbuses between Manila, Cebu and Davao in a two-class configuration seating 28 club and 232 economy passengers, but suffered from financial and operational difficulties and had to scale down its operations. Only one remains in service, with the other four returned to their lessors in spring and winter 1997.

Papua New Guinea's flag-carrier Air Niugini leased two A300B4-203s from Trans-Australia Airlines (later Australian Airlines) in the 1980s for a regional service from Auckland to Hong Kong, via the capital Port Moresby, in a short-lived tri-partite agreement with Air New Zealand and Cathay Pacific. The first A300 was delivered on 27 November 1984, and entered service at the beginning of December. It remained in service until 10 March 1989, with the second machine leased for two months between 15 December 1986 and 14 February 1987. Air Maldives has been leasing a Malaysia Airlines A300B4-203 since November 1994, flying up to 246 passengers in a two-class layout between the capital Male and Abu Dhabi, Colombo and Trivandrum.

In Africa, Airbus has done less well than in other parts of the world. Apart from Air Afrique and South African Airways, only two other airlines are operating single A300 examples on lease. One is Sudan Airways, which has been leasing an A300B4-622 on two occasions since 11 May 1994, the other is Nigerian company Bellview Airlines, operating charters out of Lagos. Bellview Airlines took delivery of an A300B4-622R on lease from Ansett Worldwide on 25 November 1997 and is using it on charters within Africa, and to such intercontinental destinations as Mumbai (Bombay) and Rio de Janeiro.

SINGAPORE INTERNATIONAL AIRLINES

Singapore Airlines joined a growing list of Asian Airbus customers on 11 May 1979, when it ordered six A300B4-203s and took out options on six more. The first was delivered on 20 December 1980. On 30 September 1981, the airline converted the six options into firm orders, and also optioned two of the smaller A310s, which was better suited to SIA's South East Asian network. Only a total of eight A300B4-203s served with the airline — four orders were cancelled in favour of the A310 — until 1985, when three were traded in to Airbus and sold to Continental Airlines, and five to Boeing. Of these, two ended up with Hapag-Lloyd, one with PIA, one with SAA and the last with Luxair.

SOUTH AFRICAN AIRWAYS

South African Airways (SAA) was an early customer for the Airbus, and the first on the African continent, when it placed an order for four A300B2Ks on 4 September 1975. The A300B2K introduced the Krüger flaps of the A300B4 to provide enhanced hot-and-high performance. SAA put the new type into service on 23 November 1976. The Airbus fleet, all powered by the CF6-50C turbofan engine, was enlarged with further orders and options taken out on 25 May 1979 and 1 July 1980, adding two A300B4-200s and one A300C4-200 convertible, the first to roll off the Toulouse production line. The A300B4-200 entered service with SAA on 28 April 1981, followed by the A300C4-200 on 31 October 1982. The last Airbus, an ex-Singapore Airlines A300B4-200 was delivered on 24 April 1985. All eight remain in frontline use, typically serving the longer domestic routes, including the main Johannesburg–Cape Town shuttle, as well as regional routes into Africa.

THAI AIRWAYS INTERNATIONAL

Thai Airways International was the first to rescue Airbus from an order draught, when it signed up for two A300B4-2Cs and two options on 15 April 1977. Deliveries took place later that year and the Airbus was placed on its regional routes on 1 November. The conversion of options and regular subsequent orders brought the fleet to eight A300B4-2C and four A300B4-203 models by the time of the final delivery on 28 March 1985. On 6 February 1981, Thai International had changed two of its options into a firm order for two of the larger A300B4-601 model, in effect launching the General Electric-powered version of the new two-crew aircraft. It came as a complete surprise to Airbus when the airline cancelled the order on 30 September 1982 in favour of two Boeing 767s, setting in train much diplomatic activity by the French Embassy in Bangkok. As a result, Thai restored the Airbus order, re-signing for two and two options on 30 April 1983, with the first joining the fleet on 26 September 1985. The Thai flag-carrier then built up its Airbus fleet on a regular basis, taking delivery of a total of six A300B4-601s, two A300B4-605s, and eight A300B4-622Rs by 5 October 1993. All but one remain in service, and five more A300B4-622Rs were delivered at the end of 1998.

ABOVE: Thai Airways International A300B4-622R.

RIGHT: South African Airways A300B2K-3C.

TRANS AUSTRALIA AIRLINES (TAA)

Australia's then government-owned domestic airline ordered four A300B4-203s and took out two options on 7 December 1979, but the vigorously fought for sale almost did not happen. Early in the new year, the Australian government threatened to block the deal in protest at the European agricultural policy, which it felt would affect the country's sheep exports. However, with TAA anxious to introduce the first wide-body aircraft onto domestic routes and score a competitive advantage over its privately-owned rival Ansett, a compromise was reached and deliveries commenced on 29 June 1981. Three had been delivered by October in time for the Australian summer season, with the fourth arriving in Australia on 30 June 1982 and a fifth — the conversion of one of the options — on 1 December 1983. The Airbus fleet was put mainly onto the eastern corridor linking Sydney, Melbourne and Brisbane, and onto the Melbourne-Perth route, configured for 230 passengers in a three-class lay-

out. All five aircraft adopted the new name of Australian Airlines on 4 August 1986. One was sold to Toa Domestic on 28 March 1987, and the rest joined the Qantas fleet when Australian Airlines was merged into the flag-carrier on 31 October 1993.

TRANS EUROPEAN AIRWAYS

Belgian-based charter operator Trans European Airways (TEA) had the distinction of operating the only A300B1 to enter airline service. The aircraft was the second test aircraft and was brought up to airline standard following TEA's order for one, plus one option, on 22 November 1974. After a short lease to Air Algérie, the A300B1 entered service with TEA after its return on 19 January 1975, equipped with 320 seats. An A300B4-2C joined the fleet on 16 October 1975, but spent most of its time out on lease, before being sold to Hapag-Lloyd in February 1982. The A300B1 was withdrawn from use in November 1990, just prior to the airline's collapse, and continues to languish at Brussels National.

TUNISAIR

Orders in Africa were hard to come by and an order from Tunisair on 5 January 1980 for one A300B4-203 and one option gave Airbus a welcome lift. The option for the second aircraft was converted into a firm order in 1983, but later cancelled.

The first wide-body aircraft to enter service with Tunisia's flag-carrier, the A300B4-203 was delivered on 28 May 1982 and put onto the Middle East and some European routes, where it still serves, fitted out for 24 business class and 241 economy class passengers. Typical destinations are Istanbul, Jeddah, Paris and Rome. Tunisair also leased an A300B4-103 between 1 April 1995 and 21 October 1995, and an ex-Air Liberté A300B4-622R for one month during summer 1997.

VARIG

Brazil's international airline took over two A300B4-203s ordered by its associate Cruzeiro do Sul, taking delivery on 3 June 1981 and 23 June 1982. Varig used the aircraft on its regional flights out of Rio de Janeiro, serving such destinations as Buenos Aires, Caracas and Miami, but such a small fleet was uneconomical to operate and the A300B4-203s were sold by the end of the decade. One went to Japan Air System on 7 December 1989, the other to Air Jamaica on 12 June 1990.

ABOVE LEFT: Tunisair A300B4-203.

LEFT: Trans Australia Airlines A300B4-203.

VASP

The second-biggest airline in Brazil announced on 3 October 1980 that it was buying three A300B2-203s for its northeastern routes out of Sao Paulo. The contract was formally signed on 19 January 1981 and the first two aircraft were flown to Brazil on 5 and 8 November 1982, with the third following on 31 January 1983. All three remain in service and are scheduled on the busiest domestic routes and times from Sao Paulo to Rio de Janeiro, Porto Alegre, Salvador and Manaus, and across the border to Buenos Aires in Argentina. They are operated with 26 first class and 214 economy class seats.

BELOW: Trans European Airways A300B1.

BOTTOM: VASP A300B2-203.

6 ACCIDENTS AND INCIDENTS

No aircraft is ever immune from the dangers of accidents, and the A300 is no exception. However, not one of the 16 hull losses to date could be traced to any technical malfunction of the airframe or systems which, in 25 years of service, is a record to be proud of and vindicates the collective design approach of the European partners. The manufacturer can do little about human fallibility, nor about the weather, and both elements appear to have had a hand in most of the Airbus crashes. Terrorism and war also touched the Airbus. An Iran Air A300B2-200 was hit by two missiles fired by the USS Vincennes after take-off from Bandar Abbas. The aircraft was apparently mistaken for an Iranian Air Force F-14 Tomcat, causing the tragic death of nearly 300 people. Two A300s on the ground found themselves in the wrong place at the wrong time. A Kuwait Airways aircraft, which had been flown to Baghdad by the Iraqis during the 1991 Gulf War, was subsequently destroyed by Allied bombing, while a second, operated by Air France, was high-jacked by four armed men at Algiers and flown to Marseille. It was damaged beyond repair when the French police stormed the aircraft.

CONTROLLED FLIGHT INTO TERRAIN

The Himalayas of Nepal witnessed a double tragedy in summer 1992, as first an Airbus A310 of Thai Airways International, and two months later a Pakistan International Airlines A300 flew into the mountainside near the capital Kathmandu. On mid-morning of 28 September, flight PK268 departed Karachi directly for Kathmandu. On board were four flight deck crew, eight cabin staff, and 151 passengers of various nationalities. The flight proceeded normally, and after making contact with Kathmandu Area Control West, the aircraft commenced its descent from Flight Level 25 (25,000ft/7,620m). At both 25 nautical miles DME and at 16nm DME, the crew reported being at 11,500ft (3,505m). When told to give its altitude at 10 nm, air traffic control was told that the aircraft was coming out

BELOW: The A300B4-203 AP-BCP belonging to Pakistan International Airlines was lost in a crash near Kathmandu on 28 September 1992. All 163 passengers and crew were killed instantly. *Christian Laugier*

of 8,500ft (2,591m). It struck the side of a ridge at the southern end of Kathmandu Valley shortly afterwards at approximately 7,350ft (2,240m). The impact was not survivable and all 163 people on board perished. Subsequent investigations revealed that the aircraft was approaching at the correct speed and attitude, with gear down, full flaps and full slats, and spoilers retracted, but, when the co-pilot reported that he was at 10,500ft (3,200m) at 16nm DME, he was actually much lower. At the point of impact, the aircraft was 1,500ft (457m) below the correct descent profile.

In September 1997, Garuda Indonesia A300-600R Flight GA 152 was on its way to Medan on the northern tip of the island of Sumatra, but crashed in a heavily wooded mountainous region near the village of Buah Nabar, killing all 234 people on board. At the time, dense smoke from hundreds of forest fires was playing havoc with air services throughout southeast Asia, and Sumatra, along with neighbouring Borneo, was one of the worst affected areas. The conditions were immediately touted as a causal factor, but excerpts published from the conversation between the flight deck and Medan control tower strongly indicated a simple human error, but with devastating consequences. Captain Rahmowiyogo, who had more than 12,000 flying hours to his credit in 20 years with the airline, was asked by the ATC controller to make a left turn, but questioned the instruction, as his knowledge of the route indicated a mountainous area to the left. 'Affirm Sir! Continue turn left on heading 215,' the controller replied. One minute before the crash, the controller asked the pilot to confirm he was making a left turn. 'We are turning right now,' the pilot replied. Even though ATC finally agreed that the pilot should turn right, it was too late. Ten seconds later, the aircraft crashed.

LOSS OF CONTROL

In what was Japan's second-worst air crash, a China Airlines A300-600R exploded and burnt out during an aborted landing in poor visibility at Nagoya on 26 April 1994. On board Flight CI 140 were 256 passengers and 15 crew members, of whom 264 were killed and seven seriously injured. According to the accident investigation report prepared the Aircraft Accident Investigation Commission under the control of Japan's Ministry of Transport, the First Officer, who was manually flying the aircraft, inadvertently activated the GO lever, which changed the flight director to GO AROUND mode and caused a thrust increase. This resulted in the aircraft deviating from its normal flight path. After the autopilots were engaged, the First Officer continued pushing the control wheel in accordance with the Captain's instructions, which caused the horizontal stabiliser to move to its full nose-up position creating an out-of-trim situation. The crew continued on its approach, apparently unaware of the abnormal situation. In the opinion of the accident investigators, the Captain, who had now taken the controls, judged that landing would be difficult and opted for a go-around. The aircraft began to climb steeply with a high pitch angle attitude and, with no effective recovery operation, stalled and crashed. The Commission suggested that with present training, the advanced automatic flight control systems are not

easily understood by crew members, and also called for manufacturers to standardise AFCS specifications.

Accident investigators must have had a feeling of *déja vu* when they pieced together what happened to another A300-600R of China Airlines, which crashed alongside Taipei's Chiang Kai Shek International Airport on the evening of 16 February 1998 as the crew was initiating a go-around. The Airbus, returning from the Indonesian island of Bali and aiming for runway 05L, crashed just beyond the western perimeter fence, clipping cars and houses before bursting into flames in a rice paddy. All 182 passengers and 14 crew lost their lives, together with six people on the ground when the aircraft hit nearby housing after impact. 'It came down . . . I heard a loud explosion and saw a fireball, and then I thought the chances for any survivors were slim,' said a local nut seller who witnessed the last moments of flight CI 676. The flight data recordings downloaded for the Taiwan Civil Aeronautics Administration (CAA) by Australia's Bureau of Air Safety Investigations, indicated that the A300-600R's Cat I ILS approach to runway 05L was far too high, forcing the crew to initiate a go-around. Although there was light rain at the time and heavy fog was reported around the airport throughout the afternoon and evening, the CAA reported that visibility had been 'adequate' at the time of the crash. It was only seconds into the manual go-around procedure after the autopilot was switched off, that the crew lost control and failed to prevent the aircraft adopting extreme pitch-up attitudes and speeds, before stalling and plunging to the ground. The distribution of the wreckage, spread over an area of 400m with only the tailfin recognisable, suggested that the aircraft hit the ground in a slight nose-up position. Although the Taiwanese CAA ordered the immediate grounding of the airline's nine A300s, no technical malfunctions were reported.

CASUALTY RECORD

The table summaries all accidents and incidents resulting in write-off of the aircraft up to 1 January 1999. This includes a total of 16, comprising one A300B2-200, four A300B4-100, four A300B4-200, one A300B4-320, one A300B4-220, two A300C4-600 and three A300B4-600R variants.

BELOW: This poor quality photograph shows the wreckage of a China Airlines A300-622R, which clipped houses before bursting into flames.

C/N	REG'N	OPERATOR	MODEL	DATE	LOCATION	TYPE OF ACCIDENT
0004	F-BUAE	Air Inter	B2-1C	31/03/93	Paris Orly	Damaged beyond repair
0022	VT-ELV	Indian Airlines	B2-1C	29/09/86	Madras	Aborted take-off and overran, tearing off undercarriage
0025	AP-BCP	PIA	B4-203	28/09/92	near Kathmandu	Crashed into mountainside near Teenpane and Bhatte Dande villages, 167 killed
0034	VT-EDV	Indian Airlines	B2-101	15/11/93	near Tirupati	Ran low on fuel and force-landed in a rice field, no casualties
0070	F-BVGK	Air France	B4-203	17/03/82	Sana'a	Engine caught fire on take-off and flames spread through aircraft, 4 injured
0072	HS-THO	Thai AI	B4-103	22/10/94	Bangkok	Damaged beyond repair
0104	F-GBEC	Air France	B2-1C	26/12/94	Marseille	Suffered major damage when stormed by French police after being hijacked at Algiers by 4 armed men
0115	SU-BCA	EgyptAir	B4-203	21/09/87	Luxor	Failed to get airborne on training flight, skidded off runway and smashed into concrete barrier, 5 killed
0122	OY-KAA	MAS	B2-320	18/12/83	Kuala Lumpur	Ended up in swamp after sixth attempt to land
0186	EP-IBU	Iran Air	B2-203	03/07/88	Arabian Gulf	Mistakenly shot down by US Navy missiles after take-off from Bandar Abbas, 290 killed
0214	PK-GAI	Garuda	B4-220	26/09/97	near Medan	Hit high ground in poor visibility. 12 crew and 222 passengers killed
0327	9K-AHF	Kuwait Airways	C4-620	15/02/91	Baghdad	Seized by Iraq on 2 August 1990 and destroyed by Allied bombing
0332	9K-AHG	Kuwait Airways	C4-620	15/02/91	Baghdad	Seized by Iraq on 2 August 1990 and destroyed by Allied bombing
0578	B-1814	China Airlines	B4-622R	16/02/98	Taipei	Crashed 2.5km short of runway in poor visibility, all 196 killed on board and 8 on ground
0580	B-1816	China Airlines	B4-622R	26/04/94	Nagoya	Crashed at the southern end of runway shortly before landing, 263 died.
0583	HL7296	Korean Air	B4-622R	10/08/94	Cheju	Overran runway while attempting to land in a tropical typhoon. Aircraft exploded and destroyed by fire. No casualties

7 PRODUCTION HISTORY

The following table provides the complete production, including the first customer delivery. Airbus test registrations, used on most aircraft, are not shown. Unless specifically indicated otherwise in the remarks column as written off (w/o), damaged beyond repair (dbr), broken up (b/u) or withdrawn from use (wfu), the aircraft remains in active service, although not necessarily with the original customer. Accidents shown did not always occur in the service of the original operator — for full details see separate casualty record. Names in italics are those first applied to individual aircraft by the first operator. Construction numbers (C/N) missing in the sequence are Airbus A310s. By 1 January 1999, Airbus had delivered a total of 481 A300 aircraft over a 30-year period, including two B1,

32 x B2-100, 25 x B2-200, 2 x B2-300, 47 x B4-100, 136 x B4-200, 3 x C4-200, 2 x F4-200, 35 x B4-600, 4 x C4-600, 164 x B4-600R, and 29 x F4-600R, plus four A300B4-600ST Beluga Super Transporters based on the A300. Individual numbers within model designations indicate the engine type, ie: B2/B4-102/202 have General Electric CF6-50C2R engines, B2/B4-1-03/2-03 the CF6-50C, B4-601/601R the CF6-80C2A1, B4-603/603R the CF6-80C2A3, B4-605/605R the CF6-80C2A5, and the B4-608ST the CF6-80C2A8. Pratt & Whitney-powered models are noted as follows: B2/B4-120/220 have JT9D-59A engines, B4-620 the JT9D-7R4H1, and the B4-622/622R the PW4158.

C/N	Reg'n	Owner/operator	Model	First Flight	Delivery	Remarks
0001	F-WUAB	Airbus Industrie	B1	28/10/72	28/10/72	w/o 27/08/74
0002	OO-TEF	Trans European Airways	B1	05/02/73	25/11/74	wfu Brussels 1/12/90
0003	F-WUAD	Airbus Industrie	B2-1C	28/06/73	28/06/73	
0004	F-BUAE	Air Inter	B2-1C	20/11/73	22/01/77	dbr Paris-Orly 31/03/93
0005	F-BVGA	Air France	B2-1C	15/04/74	10/05/74	b/u 01/08/97
0006	F-BVGB	Air France	B2-1C	23/06/74	28/06/74	
0007	F-BVGC	Air France	B2-1C	06/08/74	10/08/74	
0008	HS-VGD	Air Siam	B2-1C	02/10/74	17/10/74	*Chao Praya*
0009	HS-VGF	Air Siam	B2-1C	not taken up		
	F-ODCY	Air France	B4-103	26/12/74	16/07/76	
0010	F-BVGD	Air France	B2-1C	07/03/75	21/03/75	b/u 30/06/96
0011	F-BGVE	Air France	B2-1C	23/04/75	01/05/75	b/u 30/06/96
0012	EC-	Iberia	B4-103	not taken up		
	D-AMAX	Germanair	B4-103	20/05/75	23/05/75	*Maximilian*
0013	F-BVGF	Air France	B2-1C	31/05/75	11/06/75	b/u 15/11/96
0014	EC-	Iberia	B4-103	not taken up		
	HL7218	Korean Air Lines	B4-2C	23/07/75	01/08/75	
0015	F-BUAG	Air Inter	B2-1C	18/06/75	15/10/76	b/u 15/01/97
0016	EC-	Iberia	B4-103	not taken up		
	HL7219	Korean Air Lines	B4-2C	22/08/75	01/09/75	
0017	EC-	Iberia	B4-103	not taken up		
	OO-TEG	Trans European	B4-103	07/10/75	16/10/75	*Adrianus Andreas Jr*
0018	EC-	Iberia	B4-103	not taken up		
	HL7220	Korean Air Lines	B4-2C	23/10/75	31/10/75	
0019	FBVGG	Air France	B4-203	11/11/75	16/11/75	
0020	D-AMAY	Germanair	B4-103	19/12/75	01/04/76	*Ludwig I*
0021	D-AIAA	Lufthansa	B2-1C	29/11/75	02/02/76	*Garmish-Partenkirchen* wfu 08/11/92
0022	D-AIAB	Lufthansa	B2-1C	23/01/76	19/03/76	*Rüdesheim-am-Rhein* dbr at Madras 29/09/86
0023	F-BVGH	Air France	B4-203	12/02/76	14/04/76	
0024	HL7221	Korean Air Lines	B4-2C	03/03/76	23/04/76	
0025	D-AMAZ	Germanair	B4-103	23/03/76	02/05/77	w/o Kathmandu, Nepal 28/09/92
0026	D-AIAC	Lufthansa	B2-1C	11/03/76	25/04/76	*Lüneburg*
0027	F-BUAH	Air Inter	B2-1C	03/05/76	28/09/78	

C/N	Reg'n	Owner/operator	Model	First Flight	Delivery	Remarks
0028	HL7223	Korean Air Lines	B4-2C	16/04/76	08/07/76	
0029	HK-2057	Aerocondor Colombia	B4-102	08/05/76	10/12/77	*Ciudad de Barranquilla* w/o 03/07/97
0030	HL7224	Korean Air Lines	B4-2C	04/06/76	28/02/77	
0031	HL7238	Korean Air Lines	B4-2C	08/07/76	09/08/78	
0032	ZS-SDA	South African Airways	B2K-3C	30/07/76	15/11/76	*Blesbok*
0033	HS-TGH	Thai International	B4-103	21/09/76	25/10/77	*Srimuang*
0034	VT-EDV	Indian Airlines	B2-1C	30/08/76	31/10/76	dbr at Tirupati, India 15/11/93
0035	HS-TGK	Thai International	B4-103	02/02/77	14/12/77	*Suranaree*
0036	VT-EDW	Indian Airlines	B2-1C	05/10/76	29/11/76	
0037	ZS-SDB	South African Airways	B2K-3C	02/11/76	22/12/76	*Gemsbok*
0038	VT-EDX	Indian Airlines	B2-1C	16/11/76	29/12/76	
0039	ZS-SDC	South African Airways	B2K-3C	01/12/76	21/01/77	*Waterbok*
0040	ZS-SDD	South African Airways	B2K-3C	13/12/76	15/02/77	*Rooibok*
0041	N201EA	Eastern Air Lines	B4-103	11/01/77	03/12/77	
0042	N202EA	Eastern Air Lines	B4-103	24/02/77	19/11/77	
0043	N203EA	Eastern Air Lines	B4-103	13/05/77	29/10/77	
0044	N204EA	Eastern Air Lines	B4-103	13/07/77	24/08/77	
0045	F-BVGI	Air France	B4-203	18/02/77	24/03/77	
0046	HS-	Thai International	B4-103	not taken up		
	SX-BEB	Olympic Airways	B4-103	28/03/77	25/02/79	*Odysseus*
0047	F-BVGJ	Air France	B4-203	22/04/77	20/10/77	
0048	D-AIAD	Lufthansa	B2-1C	14/03/77	16/04/77	*Westerland-Sylt*
0049	F-ODHY	Iran Air	B2-202	10/05/77	07/03/78	*Kermanshah*
0050	F-GBEA	Air France	B2-1C	16/06/77	21/06/78	
0051	F-ODHZ	Iran Air	B2-202	03/02/78	16/03/78	*Hormozgan*
0052	D-AIAE	Lufthansa	B2-1C	17/11/77	07/01/78	*Neustadt an der Weinstrasse*
0053	D-AIBA	Lufthansa	B4-2C	16/08/77	29/09/77	*Rothenburg ob der Tauber*
0054	HS-TGL	Thai International	B4-103	05/01/78	16/02/78	*Srisoonthorn*

UPS has placed a large order for the A300F4-600 freighter.

C/N	Reg'n	Owner/operator	Model	First Flight	Delivery	Remarks
0055	HS-TGM	Thai International	B4-103	01/03/78	14/04/78	*Thepsatri*
0056	AP-	Pakistan International	B4-103	not taken up		
	SX-BEC	Olympic Airways	B4-103	19/12/78	15/02/79	*Achilleus*
0057	D-AIBB	Lufthansa	B4-2C	23/01/78	23/03/78	*Freudenstadt/Schwarzwald*
0058	RP-	Philippine Airlines	B4-103	not taken up		
	SX-BED	Olympic Airways	B4-103	19/02/80	28/03/80	*Telemachus*
0059	VT-EDY	Indian Airlines	B2-1C	29/03/78	11/05/78	
0060	VT-EDZ	Indian Airlines	B2-1C	19/03/78	08/06/78	
0061	EP-IBR	Iran Air	B2-203	21/12/79	17/03/80	
0062	F-BUAI	Air Inter	B2-1C	18/08/78	13/10/78	
0063	RP-C3001	Philippine Airlines	B4-103	10/10/79	25/11/79	
0064	D-AHLA	Hapag-Lloyd	B4-203	27/02/79	03/04/79	
0065	N205EA	Eastern Air Lines	B4-103	13/09/78	15/11/78	
0066	N206EA	Eastern Air Lines	B4-103	06/10/78	01/12/78	
0067	N207EA	Eastern Air Lines	B4-103	20/10/78	11/12/78	
0068	N208EA	Eastern Air Lines	B4-103	06/08/79	12/10/79	
0069	RP-C3002	Philippine Airlines	B4-103	23/10/79	20/12/79	
0070	F-BVGK	Air France	B4-203	22/02/79	27/04/79	W/o Sana'a, Yemen 17/03/82
0071	HS-TGN	Thai International	B4-103	17/12/78	06/03/79	*Sudawadi*
0072	HS-TGO	Thai International	B4-103	22/01/79	16/03/79	*Srichulalak*, dbr 01/07/95
0073	9M-MHA	Malaysian Airline System	B4-203	12/07/79	30/10/79	
0074	F-BVGL	Air France	B4-203	27/03/79	15/05/79	
0075	D-AIBC	Lufthansa	B4-2C	31/01/79	23/03/79	*Lindau/Bodensee*
0076	D-AIBD	Lufthansa	B4-203	19/02/79	30/03/79	*Erbach/Odenwald*
0077	D-AIBF	Condor Flugdienst	B4-203	08/03/79	19/04/79	
0078	F-BVGM	Air France	B4-203	06/04/79	31/05/79	
0079	LN-RCA	SAS	B2-320	28/04/79	15/01/80	*Snorre Viking*
0080	EP-IBS	Iran Air	B2-203	13/02/80	16/04/80	
0081	HL7246	Korean Air Lines	B4-2C	09/05/79	29/06/79	
0082	JA8464	Toa Domestic Airlines	B2K-3C	18/08/80	02/10/80	
0083	D-AHLB	Hapag-Lloyd	C4-203	16/05/79	31/01/80	
0084	HS-TGP	Thai International	B4-103	08/06/79	09/08/79	*Srisubhan*
0085	HS-TGR	Thai International	B4-103	03/07/79	06/09/79	*Thepamart*
0086	N209EA	Eastern Air Lines	B4-103	26/07/79	14/11/79	
0087	N210EA	Eastern Air Lines	B4-103	22/08/79	15/11/79	
0088	VT-EFV	Indian Airlines	B2-1C	03/09/79	25/10/79	
0089	JA8465	Toa Domestic Airlines	B2K-3C	28/10/80	15/12/80	
0090	JA8466	Toa Domestic Airlines	B2K-3C	05/11/80	18/12/80	
0091	N212EA	Eastern Air Lines	B4-103	26/09/79	30/11/79	
0092	N213EA	Eastern Air Lines	B4-103	17/10/79	10/12/79	
0093	9M-MHB	Malaysian Airline System	B4-203	18/10/79	28/12/79	
0094	SE-DFK	SAS	B2-320	06/12/79	10/03/80	*Sven Viking*
0095	9M-MHC	Malaysian Airline System	B4-203	20/11/79	18/01/80	
0096	AP-BAX	Pakistan International	B4-203	30/11/79	03/03/80	
0097	F-BUAJ	Air Inter	B2-1C	17/01/80	28/02/80	
0098	AP-BAY	Pakistan International	B4-203	25/01/80	12/03/80	
0099	AP-BAZ	Pakistan International	B4-203	22/02/80	03/04/80	
0100	F-BVGN	Air France	B4-203	21/01/80	06/03/80	
0101	I-BUSB	Alitalia	B4-203	30/01/80	28/04/80	*Tiziano*
0102	F-GBEB	Air France	B2-1C	08/02/80	20/03/80	
0103	SX-BEE	Olympic Airways	B4-103	06/03/80	21/04/80	*Nestor*
0104	F-GBEC	Air France	B2-1C	28/02/80	08/04/80	dbr at Marseille, France 26/12/94
0105	SX-BEF	Olympic Airways	B4-103	14/03/80	30/04/80	*Ajax*
0106	I-BUSC	Alitalia	B4-203	21/03/80	29/05/80	*Botticelli*
0107	I-BUSD	Alitalia	B4-203	31/03/80	20/06/80	*Caravaggio*
0108	N215EA	Eastern Air Lines	B4-103	11/04/80	10/06/80	
0109	PP-CLA	Cruzeiro do Sul	B4-203	23/04/80	20/06/80	
0110	PP-CLB	Cruzeiro do Sul	B4-203	06/05/80	26/06/80	
0111	VT-EFW	Indian Airlines	B2-1C	19/05/80	18/07/80	
0112	F-BUAK	Air Inter	B2K-3C	24/09/80	25/11/80	wfu and st Chateauroux, France
0113	VT-EFX	Indian Airlines	B2-1C	10/06/80	07/08/80	

C/N	Reg'n	Owner/operator	Model	First Flight	Delivery	Remarks
0114	AP-BBA	Pakistan International	B4-203	16/06/80	27/08/80	
0115	SU-BCA	EgyptAir	B4-203	03/07/80	19/09/80	*Horus*, dbr at Luxor, Egypt 21/09/87
0116	SU-BCB	EgyptAir	B4-203	10/07/80	30/09/80	*Osiris*
0117	9V-STA	Singapore Airlines	B4-203	19/08/80	20/12/80	
0118	N216EA	Eastern Air Lines	B4-103	26/08/80	13/10/80	
0119	N217EA	Eastern Air Lines	B4-103	04/09/80	22/10/80	b/u 01/01/97
0120	N219EA	Eastern Air Lines	B4-103	15/09/80	01/12/80	
0121	9V-STB	Singapore Airlines	B4-203	24/11/80	24/02/81	
0122	OY-KAA	SAS	B2-120	07/10/80	12/12/80	*Stig Viking*, dbr at Kuala Lumpur 18/12/83
0123	I-BUSF	Alitalia	B4-203	10/10/80	02/12/80	*Tintoretto*
0124	N220EA	Eastern Air Lines	B4-103	22/10/80	10/12/80	
0125	RP-C3003	Philippine Airlines	B4-203	31/10/80	10/04/81	
0126	9V-STC	Singapore Airlines	B4-203	29/01/81	20/03/81	
0127	G-BIMA	Laker Airways	B4-203	14/11/80	07/01/81	*Metro*
0128	SE-DFL	SAS	B2-120	11/12/80	12/03/81	*Ingemar Viking*
0129	F-BVGO	Air France	B4-203	05/12/80	05/02/81	
0130	EC-DLE	Iberia	B4-120	13/01/81	27/02/81	*Doana*
0131	G-BIMB	Laker Airways	B4-203	30/12/80	17/02/81	*Orient Express*
0132	D-AIAF	Lufthansa	B2-1C	08/01/81	19/02/81	
0133	EC-DLF	Iberia	B4-120	10/12/80	20/03/81	*Canadas del Teide*
0134	VH-TAA	Trans-Australia Airlines	B4-203	06/05/81	29/06/81	*James Cook*
0135	EC-DLG	Iberia	B4-120	25/02/81	03/04/81	*Tablas de Daimiel*
0136	EC-DLH	Iberia	B4-120	07/03/81	14/04/81	*Aigues Tortes*
0137	TU-TAO	Air Afrique	B4-203	20/02/81	07/05/81	*Nouakchott*
0138	ZS-SDE	South African Airways	B4-203	03/03/81	23/04/81	*Springbok*
0139	I-BUSG	Alitalia	B4-203	11/03/81	28/04/81	*Canaletto*
0140	I-BUSH	Alitalia	B4-203	24/03/81	13/05/81	*Mantegna*
0141	HS-TGT	Thai International	B4-203	31/03/81	15/05/81	*Jiraprabha*
0142	I-BUSJ	Alitalia	B4-203	03/04/81	27/05/81	*Tiepolo*
0143	PP-VND	Varig	B4-203	08/04/81	03/06/81	
0144	G-BIMC	Laker Airways	B4-203	16/04/81	11/06/81	*Intercity Express*
0145	F-BVGP	Air France	B4-203	23/04/81	15/06/81	
0146	F-BVGQ	Air France	B4-203	04/05/81	26/06/81	
0147	9M-MHD	Malaysian Airline System	B4-203	11/05/81	24/07/81	
0148	SX-BEG	Olympic Airways	B4-103	19/05/81	29/07/81	*Diomedes*
0149	HS-TGW	Thai International	B4-203	27/05/81	30/09/81	*Srisachanalai*
0150	SU-BCC	EgyptAir	B4-203	10/06/81	11/08/81	*Nout*
0151	VH-TAB	Trans-Australia Airlines	B4-203	09/06/81	18/08/81	*John Oxley*
0152	N221EA	Eastern Air Lines	B4-203	22/06/81	07/10/81	
0153	N222EA	Eastern Air Lines	B4-203	30/06/81	15/10/81	
0154	N223EA	Eastern Air Lines	B4-203	07/07/81	09/11/81	
0155	N224EA	Eastern Air Lines	B4-203	12/08/81	09/10/81	
0156	EC-DNQ	Iberia	B4-120	28/10/81	02/02/82	*Islas Cies*
0157	VH-TAC	Trans-Australia Airlines	B4-203	24/08/81	09/10/81	*John Forrest*
0158	N225EA	Eastern Air Lines	B4-203	03/09/81	24/11/81	
0159	PK-GAA	Garuda Indonesia	B4-220	06/10/81	04/03/82	
0160	JA8471	Toa Domestic Airlines	B2K-3C	17/09/81	13/11/81	
0161	N226EA	Eastern Air Lines	B4-203	24/09/81	04/12/81	
0163	JA8472	Toa Domestic Airlines	B2K-3C	06/10/81	08/12/81	
0164	PK-GAC	Garuda Indonesia	B4-220	03/11/81	25/02/82	
0165	PK-GAD	Garuda Indonesia	B4-220	12/11/81	11/01/82	
0166	PK-GAE	Garuda Indonesia	B4-220	25/11/81	19/01/82	
0167	PK-GAF	Garuda Indonesia	B4-220	10/12/81	03/02/82	
0168	PK-GAG	Garuda Indonesia	B4-220	30/12/81	10/02/82	
0169	9V-STD	Singapore Airlines	B4-203	18/11/81	12/02/82	
0170	EC-DNR	Iberia	B4-120	13/01/82	26/02/82	*Ordesa*
0171	EC-DNS	Iberia		not taken up		
	B-1812	China Airlines	B4-220	20/04/82	04/06/87	
0173	I-BUSL	Alitalia	B4-203	14/12/81	23/02/82	*Pinturicchio*

C/N	Reg'n	Owner/operator	Model	First Flight	Delivery	Remarks
0174	9V-STE	Singapore Airlines	B4-203	15/12/81	22/02/82	
0175	F-BVGR	Air France	B4-203	02/02/82	07/04/82	
0176	JA8473	Toa Domestic Airlines	B2K-3C	06/01/82	17/02/82	
0177	G-BIMD	Laker Airways	B4-203	not taken up		
	VT-EHN	Air-India	B4-203	15/01/82	28/07/82	*Ganga*
0178	F-BVGS	Air France	B4-203	21/01/82	10/03/82	
0179	EC-DNT	Iberia	B4-220	not taken up		
	B-1810	China Airlines	B4-220	05/05/82	05/07/85	
0180	G-BIME	Laker Airways	B4-203	not taken up		
	VT-EHO	Air-India	B4-203	17/06/82	10/08/82	*Godavari*
0181	VT-EHC	Indian Airlines	B4-203	18/03/82	24/05/82	
0182	VT-EHD	Indian Airlines	B4-203	26/03/82	27/05/82	
0183	F-BVGT	Air France	B4-203	25/02/82	15/04/82	
0184	SX-BEH	Olympic Airways	B4-103	06/03/82	08/04/82	*Peleus*
0185	EP-IBT	Iran Air	B2-203	09/03/82	30/04/82	
0186	EP-IBU	Iran Air	B2-203	16/03/82	30/04/82	dbr 03/07/88
0187	EP-IBV	Iran Air	B2-203	23/03/82	12/05/82	
0188	TS-IMA	Tunis Air	B4-203	13/04/82	28/05/82	*Amilcar*
0189	SX-BEI	Olympic Airways	B4-103	02/04/82	17/05/82	*Neoptolemos*
0190	G-BIMF	Laker Airways	B4-203	not taken up		
	VT-EHQ	Air-India	B4-203	02/08/82	05/11/82	*Cauvery*
0192	ZS-SDF	South African Airways	B4-203	20/04/82	09/06/82	*Eland*
0193	B-190	China Airlines	B4-220	28/04/82	22/06/82	
0194	PP-VNE	Varig	B4-203	04/05/80	23/06/80	
0195	5A-	Libyan Arab Airlines	B4-203	not taken up		
	N202PA	Pan American	B4-203	16/09/82	21/12/84	*Clipper America*
0196	VH-TAD	Trans-Australia Airlines	B4-203	13/05/82	30/06/82	*William Light*
0197	B-192	China Airlines	B4-220	01/06/82	23/07/82	
0198	5A-	Libyan Arab Airlines	B4-203	not taken up		
	N204PA	Pan American	B4-203	08/10/82	21/12/84	*Clipper Costa Rica*
0199	SU-BDF	EgyptAir	B4-203	09/06/82	16/08/82	*Hathor*
0200	SU-BDG	EgyptAir	B4-203	17/06/82	25/08/82	*Atoum*
0202	PP-SNL	VASP	B2-203	28/06/82	05/11/82	
0203	RP-C3004	Philippine Airlines	B4-203	01/07/82	29/03/83	
0204	N227EA	Eastern Airlines	B4-203	29/07/82	14/10/82	
0205	PP-SNM	VASP	B2-203	13/07/92	05/11/82	
0206	-			aircraft not built		
0207	N228EA	Eastern Air Lines	B4-203	20/08/82	21/12/82	
0208	6Y-	Air Jamaica	B4-203	not taken up		
	N212PA	Pan American	B4-203	23/08/82	17/05/85	*Clipper Detroit*
0209	JA8476	Toa Domestic Airlines	B2K-3C	11/10/82	28/02/83	
0210	6Y-	Air Jamaica	B4-203	not taken up		
	N213PA	Pan American	B4-203	10/09/82	10/05/85	*Clipper Chicago*
0211	N229EA	Eastern Air Lines	B4-203	15/09/82	22/12/82	
0212	ZS-SDG	South African Airways	C4-203	17/09/82	29/10/82	*Koedoe*
0213	PK-GAH	Garuda Indonesia	B4-220	29/09/82	10/11/82	
0214	PK-GAI	Garuda Indonesia	B4-220	06/10/82	16/11/82	dbr near Medan 26/09/97
0215	PK-GAJ	Garuda Indonesia	B4-220	13/10/82	23/11/82	
0216	N230EA	Eastern Air Lines	B4-203	18/10/82	22/12/82	
0218	VH-TAE	Trans-Australia Airlines	B4-203	29/10/82	01/12/83	*John Fawkner*
0219	RP-C3005	Philippine Airlines	B4-203	04/11/82	29/03/83	
0220	N231EA	Eastern Air Lines	B4-203	05/11/82	29/12/82	
0221	B-194	China Airlines	B4-220	18/11/82	21/12/82	
0222	9V-STF	Singapore Airlines	B4-203	22/11/82	23/12/82	
0223	-			aircraft not built		
0225	PP-SNN	VASP	B2-203	16/12/82	31/01/83	
0226	EP-IBZ	Iran Air	B2-203	13/12/82	31/01/83	
0227	N203PA	Pan American	B4-203	01/03/83	21/12/84	*Clipper New York*
0228	-			aircraft not built		
0229	G-BIMG	Laker Airways		not taken up	aircraft not built	
0231	-			aircraft not built		
0232	B-196	China Airlines	B4-220	03/01/83	27/07/83	

C/N	Reg'n	Owner/operator	Model	First Flight	Delivery	Remarks
0234	SX-	Olympic Airways	B4-103	not taken up		
	N206PA	Pan American	B4-203	18/07/83	25/04/85	Clipper Tampa
0235	I-	Alitalia	B4-203	not taken up		
	N211PA	Pan American	B4-203	18/01/83	26/04/85	*Clipper Houston*
0236	SX-	Olympic Airways	B4-103	not taken up		
	N207PA	Pan American	B4-203	18/10/83	18/03/85	*Clipper Panama*
0238	G-BIMH	Laker Airways	B4-203	not taken up		
	N210PA	Pan American	B4-203	24/02/83	25/04/85	*Clipper Dallas*
0239	SU-GAA	EgyptAir	B4-203	14/02/83	11/04/83	*Isis*
0240	SU-GAB	EgyptAir	B4-203	18/02/83	03/05/83	*Amun*
0242	-			aircraft not built		
0243	TU-TAS	Air Afrique	B4-203	04/03/83	12/07/83	
0244	JA8477	Toa Domestic Airlines	B2K-3C	09/03/83	22/04/83	
0246	-			aircraft not built		
0247	5A-	Libyan Arab Airlines	B4-203	not taken up		
	N205PA	Pan American	B4-203	09/09/83	21/12/84	*Clipper Miami*
0249	HS-TGX	Thai International	B4-203	23/06/83	08/03/85	*Srisoryothai*
0250	VT-	Indian Airlines	B4-203	not taken up		
	N970C	Continental Airlines	B4-203	31/05/83	25/04/86	
0252	F-ODTK	La Tur Charter Airlines	B4-622	08/07/83	03/08/89	*Yum'ik*
0253	JA8478	Toa Domestic Airlines	B2K-3C	04/05/83	16/06/83	
0255	SU-GAC	EgyptAir	B4-203	11/05/83	04/07/83	*Bennou*
0256	5A-	Libyan Arab Airlines	B4-203	not taken up		
	JA8237	Toa Domestic Airlines	C4-203	18/05/83	25/03/86	
0258	-			aircraft not built		
0259	N232EA	Eastern Air Lines	B4-203	17/08/83	16/12/83	
0261	N233EA	Eastern Air Lines	B4-203	25/08/83	16/12/83	
0262	N971C	Continental Airlines	B4-203	17/11/83	06/05/86	
0263	-			aircraft not built		
0265	HS-TGY	Thai International	B4-203	10/08/83	28/03/85	*Pathoomawadi*

BELOW: **Vietnam Airlines A300B4-103**. *Christian Laugier*

C/N	Reg'n	Owner/operator	Model	First Flight	Delivery	Remarks
0266	-			aircraft not built		
0268	9V-STG	Singapore Airlines	B4-203	27/07/83	09/09/83	
0269	9V-STH	Singapore Airlines	B4-203	12/08/83	22/09/83	
0271	N234EA	Eastern Air Lines	B4-203	01/09/83	19/12/83	
0272	-			aircraft not built		
0274	N235EA	Eastern Air Lines	B4-203	20/09/83	19/12/83	
0275	PK-	Garuda Indonesia	B4-220	order cancelled	aircraft not built	
0277	5A-	Libyan Arab Airlines	B4-203	not taken up		
	HL7278	Korean Air	F4-203	29/09/83	05/08/86	
0279	-			aircraft not built		
0282	G-BIMI	Laker Airways	B4-203	not taken up		
	TU-TAT	Air Afrique	B4-203	13/10/83	13/09/84	
0284	HZ-AJA	Saudia	B4-620	15/11/83	01/06/84	
0286	PK-	Garuda Indonesia	B4-220	order cancelled	aircraft not built	
0287	VT-	Indian Airlines	B4-203	order cancelled	aircraft not built	
0289	N972C	Continental Airlines	B4-203	22/11/83	25/04/86	
0290	-			aircraft not built		
0292	HL7279	Korean Air	F4-203	31/10/84	06/08/86	
0294	HZ-AJB	Saudia	B4-620	20/12/83	09/04/84	
0296	TS-	Tunis Air	B4-203	order cancelled	aircraft not built	
0298	-			aircraft not built		
0299	OH-LAA	Kar-Air	B4-203	09/08/84	12/12/86	
0301	HZ-AJC	Saudia	B4-620	03/01/84	25/03/84	
0302	OH-LAB	Kar-Air	B4-203	14/11/84	13/03/87	
0304	9V-	Singapore Airlines	B4-203	not taken up		
	N208PA	Pan American	B4-203	02/01/85	25/04/85	*Clipper San Francisco*
0305	G-BIMJ	Laker Airways	B4-203	not taken up		
	N209PA	Pan American	B4-203	30/12/84	29/03/85	*Clipper Guatemala*
0307	HZ-AJD	Saudia	B4-620	02/02/84	03/04/84	
0308	9V-STI	Singapore Airlines	B4-203	order cancelled	aircraft not built	
0310	9V-STJ	Singapore Airlines	B4-203	order cancelled	aircraft not built	
0312	HZ-AJE	Saudia	B4-620	20/02/84	20/04/84	
0314	AP-	Pakistan International	B4-203	order cancelled	aircraft not built	
0315	AP-	Pakistan International	B4-203	order cancelled	aircraft not built	
0317	HZ-AJF	Saudia	B4-620	07/03/84	28/04/84	
0319	-			aircraft not built		
0321	HZ-AJG	Saudia	B4-620	29/03/84	14/05/84	
0322	-			aircraft not built		
0324	-			aircraft not built		
0325	-			aircraft not built		
0327	9K-AHF	Kuwait Airways	C4-620	12/04/84	30/05/84	dbr at Baghdad, Iraq 15/02/91
0328	-			aircraft not built		
0330	-			aircraft not built		
0332	9K-AHG	Kuwait Airways	C4-620	26/04/84	08/06/84	
0334	-			aircraft not built		
0336	HZ-AJH	Saudia	B4-620	29/05/84	04/07/84	
0337	-			aircraft not built		
0341	HZ-AJI	Saudia	B4-620	28/06/84	18/08/84	
0344 France	9K-AHI	Kuwait Airways	C4-620	16/06/84	31/08/84	wfu and stored at Toulouse,
0348	HZ-AJJ	Saudia	B4-620	10/07/84	27/08/84	
0351	HZ-AJK	Saudia	B4-620	23/07/84	08/10/84	
0354	A6-SHZ	Abu Dhabi Government	B4-620	10/12/84	30/09/85	
0358	HL7287	Korean Air	B4-620	05/03/86	11/03/88	
0361	HL7280	Korean Air	B4-620	05/09/86	04/08/87	
0365	HL7281	Korean Air	B4-620	25/03/87	31/08/87	
0366	HZ-	Saudia	C4-620	order cancelled	aircraft not built	
0368	HS-TAA	Thai International	B4-601	01/07/85	26/09/85	*Suwannaphum*
0371	HS-TAB	Thai International	B4-601	12/07/85	27/09/85	*Sri Anocha*
0374	A6-PFD	Abu Dhabi Government	C4-620	02/10/85	05/12/85	
0377	HS-TAC	Thai International	B4-601	16/09/85	06/12/85	*Sri Ayutthaya*
0380	D-AIAH	Lufthansa	B4-603	31/12/86	12/03/87	*Lindau/Bodensee*

C/N	Reg'n	Owner/operator	Model	First Flight	Delivery	Remarks
0384	HS-TAD	Thai International	B4-601	27/11/85	03/02/86	*Uthong*
0388	7T-	Air Algerié	B4-620	not taken up		
	HL7290	Korean Air	B4-622	28/10/86	23/03/89	
0391	D-AIAI	Lufthansa	B4-603	08/01/87	26/03/87	*Erbach/Odenwald*
0395	HS-TAE	Thai International	B4-601	13/08/86	09/10/86	*Sukhothai*
0398	HS-TAF	Thai International	B4-601	25/09/86	18/12/86	*Ratchasima*
0401	D-AIAK	Lufthansa	B4-603	02/02/87	02/04/87	*Kronburg im Taunus*
0405	D-AIAL	Lufthansa	B4-603	03/02/87	18/04/87	*Stade*
0408	D-AIAM	Lufthansa	B4-603	23/02/87	30/04/87	*Rosenheim*
0411	D-AIAN	Lufthansa	B4-603	27/03/87	05/06/87	*Nördlingen*
0414	D-AIAP	Lufthansa	B4-603	17/04/87	25/06/87	*Bingen*
0417	HL7291	Korean Air	B4-622	01/06/87	28/04/89	
0420	N14053	American Airlines	B4-605R	09/12/87	12/07/88	
0423	N91050	American Airlines	B4-605R	10/02/88	20/04/88	
0459	N50051	American Airlines	B4-605R	04/03/88	15/05/88	
0460	N80052	American Airlines	B4-605R	22/03/88	31/05/88	
0461	N70054	American Airlines	B4-605R	14/04/88	23/06/88	
0462	N7055A	American Airlines	B4-605R	18/03/88	10/06/88	
0463	N14056	American Airlines	B4-605R	25/04/88	29/06/88	
0464	HS-TAG	Thai International	B4-605R	17/05/88	04/08/88	*Srinapha*
0465	N80057	American Airlines	B4-605R	24/05/88	31/08/88	
0466	N80058	American Airlines	B4-605R	23/06/88	09/09/88	
0469	N19059	American Airlines	B4-605R	17/06/88	19/09/88	
0470	N11060	American Airlines	B4-605R	07/07/88	13/10/88	
0471	N14061	American Airlines	B4-605R	23/08/88	25/10/88	
0474	N7062A	American Airlines	B4-605R	08/09/88	15/11/88	
0477	HL7288	Korean Air	B4-622R	03/10/88	29/11/88	
0479	HL7289	Korean Air	B4-622R	20/09/88	13/12/88	
0505	A6-EKC	Emirates Airline	B4-605R	19/10/88	16/05/89	

BELOW: The White Whale.

C/N	Reg'n	Owner/operator	Model	First Flight	Delivery	Remarks
0506	N41063	American Airlines	B4-605R	16/11/88	02/02/89	
0507	N40064	American Airlines	B4-605R	23/11/88	15/02/89	
0508	N14065	American Airlines	B4-605R	06/12/88	23/02/89	
0509	N18066	American Airlines	B4-605R	22/12/88	28/02/89	
0510	N8067A	American Airlines	B4-605R	12/01/89	23/03/89	
0511	N14068	American Airlines	B4-605R	24/01/89	06/04/89	
0512	N33069	American Airlines	B4-605R	20/02/89	20/04/89	
0513	N90070	American Airlines	B4-605R	01/03/89	26/04/89	
0514	N25071	American Airlines	B4-605R	23/03/89	23/05/89	
0515	N70072	American Airlines	B4-605R	01/04/89	01/06/89	
0516	N70073	American Airlines	B4-605R	01/04/89	15/06/89	
0517	N70074	American Airlines	B4-605R	24/04/89	28/06/89	
0518	HS-TAH	Thai International	B4-605R	12/05/89	30/06/89	*Napachinda*
0521	B-2306	China Eastern Airlines	B4-605R	02/06/89	24/11/89	
0525	B-2307	China Eastern Airlines	B4-605R	19/07/89	18/12/89	
0529	B-1814	China Airlines	B4-622R	not taken up		
	B-1800	China Airlines	B4-622R	10/07/89	25/09/89	
0530	F-ODSX	La Tur Charter Airlines	B4-622R	25/07/89	13/12/89	*Mexico*
0532	B-2308	China Eastern Airlines	B4-605R	06/09/89	01/12/89	
0533	B-1816	China Airlines	B4-622R	not taken up		
	B-1802	China Airlines	B4-622R	15/09/89	17/11/89	
0536	B-1818	China Airlines	B4-622R	not taken up		
	B-1804	China Airlines	B4-622R	11/10/89	15/12/89	
0540	G-MONR	Monarch Airlines	B4-605R	28/09/89	15/03/90	
0543	HL7292	Korean Air	B4-622R	09/11/89	16/01/90	
0546	D-AIAR	Lufthansa	B4-603	06/11/89	02/02/90	*Bingen*
0553	D-AIAS	Lufthansa	B4-603	08/12/89	06/03/90	*Mönchengladbach*
0554	HL7293	Korean Air	B4-622R	11/01/90	12/03/90	
0555	F-GHEF	Air Liberté	B4-622R	12/01/90	22/03/90	
0556	G-MONS	Monarch Airlines	B4-605R	14/02/90	17/04/90	
0557	SU-GAR	EgyptAir	B4-622R	27/02/90	29/05/90	*Zoser*
0558	A6-EKD	Emirates Airline	B4-605R	20/03/90	11/06/90	
0559	F-GHEG	Air Liberté	B4-622R	12/04/90	07/06/90	
0560	HL7294	Korean Air	B4-622R	25/04/90	29/06/90	
0561	SU-GAS	EgyptAir	B4-622R	25/06/90	25/08/90	
0563	A6-EKE	Emirates Airline	B4-605R	11/07/90	25/09/90	
0566	HS-TAK	Thai International	B4-622R	14/08/90	16/10/90	*Phaya Thai*
0569	HS-TAL	Thai International	B4-622R	04/09/90	09/11/90	*Sritrang*
0572	SU-GAT	EgyptAir	B4-622R	13/09/90	20/11/90	*Chephren*
0575	SU-GAU	EgyptAir	B4-622R	19/09/90	10/12/90	*Mycerinus*
0577	HS-TAM	Thai International	B4-622R	02/10/90	04/12/90	*Chiang Mai*
0578	B-1814	China Airlines	B4-622R	16/10/90	14/12/90	dbr at Taipei 16/02/98
0579	SU-GAV	EgyptAir	B4-622R	25/10/90	11/02/91	*Menes*
0580	B-1816	China Airlines	B4-622R	30/10/90	28/01/91	dbr at Nagoya, Japan 26/04/94
0581	SU-GAW	EgyptAir	B4-622R	16/11/90	19/07/91	*Ahmuse*
0582	HL7295	Korean Air	B4-622R	29/11/90	26/02/91	
0583	HL7296	Korean Air	B4-622R	06/12/90	26/02/91	dbr at Cheju 10/08/94
0584	VH-YMA	Compass Airlines	B4-605R	14/12/90	03/04/91	
0601	SU-GAX	EgyptAir	B4-622R	11/01/91	27/08/91	*Tut-Ankh-Amun*
0602	JA8375	Japan Air System	B4-622R	23/01/91	25/04/91	
0603	VH-YMB	Compass Airlines	B4-605R	30/01/91	19/08/91	
0604	G-MONT	Monarch Airlines	B4-605R	not taken up		
	G-MAJS	Monarch Airlines	B4-605R	13/02/91	26/04/91	
0605	G-MONU	Monarch Airlines	B4-605R	not taken up		
	G-OJMR	Monarch Airlines	B4-605R	26/02/91	03/05/91	
0606	N3075A	American Airlines	B4-605R	08/03/91	23/05/91	
0607	SU-GAY	EgyptAir	B4-622R	26/03/91	05/09/91	*Seti I*
0608	A6-EKF	Emirates Airline	B4-605R	10/04/91	14/06/91	
0609	HL7297	Korean Air	B4-622R	19/04/91	03/07/91	
0610	N7076A	American Airlines	B4-605R	06/05/91	11/07/91	
0611	PK-GAK	Garuda Indonesia	B4-622R	24/05/91	22/08/91	
0612	N14077	American Airlines	B4-605R	05/06/91	29/08/91	

C/N	Reg'n	Owner/operator	Model	First Flight	Delivery	Remarks
0613	PK-GAL	Garuda Indonesia	B4-622R	14/06/91	24/09/91	
0614	HL7298	Korean Air	B4-622R	27/06/91	20/09/91	
0615	N34078	American Airlines	B4-605R	08/07/91	27/09/91	
0616	SU-GAZ	EgyptAir	B4-622R	21/08/91	16/12/91	
0617	JA8376	Japan Air System	B4-622R	22/08/91	18/11/91	
0618	D-AIAT	Lufthansa	B4-603	04/09/91	19/11/91	*Bottrop*
0619	N70079	American Airlines	B4-605R	18/09/91	09/12/91	
0621	JA8377	Japan Air Ssytem	B4-622R	02/10/91	27/01/92	
0623	D-AIAU	Lufthansa	B4-603	23/10/91	13/01/92	*Bocholt*
0625	PK-GAM	Garuda Indonesia	B4-622R	18/11/91	31/03/92	
0626	N77080	American Airlines	B4-605R	03/12/91	08/04/92	
0627	HL7239	Korean Air	B4-622R	07/01/92	26/03/92	
0628	HS-TAN	Thai International	B4-622R	16/01/92	07/04/92	*Chiang Rai*
0629	HS-TAO	Thai International	B4-622R	29/01/92	21/04/92	*Chanthaburi*
0630	PK-GAN	Garuda Indonesia	B4-622R	12/02/92	22/04/92	
0631	HL7240	Korean Air	B4-622R	26/02/92	19/05/92	
0632	SX-BEX	Olympic Airways	B4-605R	11/03/92	04/06/92	*Macedonia*
0633	PK-GAO	Garuda Indonesia	B4-622R	01/04/92	30/06/92	
0635	HS-TAP	Thai International	B4-622R	09/04/92	29/06/92	*Pathum Thani*
0637	JA8558	Japan Air System	B4-622R	22/04/92	28/07/92	
0639	N59081	American Airlines	B4-605R	05/05/92	24/07/92	
0641	JA8559	Japan Air System	B4-622R	14/05/92	24/09/92	
0643	N7082A	American Airlines	B4-605R	09/06/92	11/09/92	
0645	N7083A	American Airlines	B4-605R	17/06/92	02/10/92	
0655	F-GSTA	Airbus Inter Transport	B4-608ST	13/09/94	25/10/95	
0657	PK-GAP	Garuda Indonesia	B4-622R	22/07/92	30/09/92	
0659	PK-GAQ	Garuda Indonesia	B4-622R	27/07/92	29/10/92	
0662	HL7241	Korean Air	B4-622R	21/08/92	12/01/93	
0664	PK-GAR	Garuda Indonesia	B4-622R	10/09/92	01/12/92	

BELOW: The Abu Dhabi Government's single A300B4-620 at Le Bourget, Paris in June 1998. *Günter Endres*

C/N	Reg'n	Owner/operator	Model	First Flight	Delivery	Remarks
0666	B-1806	China Airlines	B4-622R	25/09/92	10/12/92	
0668	PK-GAS	Garuda Indonesia	B4-622R	27/10/92	17/12/92	
0670	JA8561	Japan Air System	B4-622R	27/10/92	21/01/93	
0673	9K-AMA	Kuwait Airways	B4-605R	10/11/92	28/05/93	*Failaka*
0675	N80084	American Airlines	B4-605R	25/11/92	19/02/93	
0677	PK-GAT	Garuda Indonesia	B4-622R	03/12/92	02/03/93	
0679	JA8562	Japan Air System	B4-622R	17/12/92	08/03/93	
0681	HS-TAT	Thai International	B4-622R	04/01/93	18/03/93	*Yasothon*
0683	JA8563	Japan Air System	B4-622R	27/01/93	16/06/93	
0685	HL7242	Korean Air	B4-622R	25/01/93	08/04/93	
0688	PK-	Garuda Indonesia	B4-622R	not taken up		
	B-2311	China Northern Airlines	B4-622R	11/02/93	15/06/93	
0690	PK-	Garuda Indonesia	B4-622R	not taken up		
	B-2312	China Northern Airlines	B4-622R	11/03/93	21/06/93	
0692	HL7243	Korean Air	B4-622R	29/03/93	29/06/93	
0694	9K-AMB	Kuwait Airways	B4-605R	07/04/93	08/07/93	*Burghan*
0696	SX-BEL	Olympic Airways	B4-605R	26/04/93	04/10/93	*Athena*
0699	9K-AMC	Kuwait Airways	B4-605R	21/04/93	22/07/93	*Wafra*
0701	A6-EKM	Emirates Airline	B4-605R	17/05/93	31/08/93	
0703	JA8564	Japan Air System	B4-622R	17/06/93	06/12/94	
0705	HS-TAS	Thai International	B4-622R	12/07/93	05/10/93	*Yala*
0707	B-2318	China Eastern Airlines	B4-605R	16/07/93	22/10/93	
0709	B-2320	China Eastern Airlines	B4-605R	02/09/93	14/04/94	
0711	JA8565	Japan Air System	B4-622R	20/09/93	15/12/93	
0713	B-2321	China Eastern Airlines	B4-622R	07/10/93	03/05/94	
0715	B-2322	China Eastern Airlines	B4-605R	07/10/93	20/04/94	
0717	HL7299	Korean Air	B4-622R	21/10/93	28/01/94	
0719	9K-AMD	Kuwait Airways	B4-605R	05/11/93	21/01/94	
0721	9K-AME	Kuwait Airways	B4-605R	18/11/93	01/02/94	*Al-Rawdhatain*
0722	HL7244	Korean Air	B4-605R	02/12/93	16/03/94	
0723	EP-IBA	Iran Air	B4-605R	21/12/93	27/12/94	
0724	JA8527	Japan Air System	B4-622R	11/01/94	11/04/95	
0725	B-2324	China Northwest Airlines	B4-605R	03/08/94	26/10/94	
0726	N650FE	Federal Express	F4-605R	02/12/93	24/05/94	*Molly Mickler*
0727	EP-IBB	Iran Air	B4-605R	18/01/94	27/12/94	
0728	N651FE	Federal Express	F4-605R	28/01/94	28/04/94	*Diane Kathleen*
0729	JA8529	Japan Air System	B4-622R	16/02/94	31/05/94	
0730	JA8566	Japan Air System	B4-622R	28/02/94	01/09/95	
0731	HL7245	Korean Air	B4-622R	15/03/94	26/05/94	
0732	B-2319	China Eastern Airlines	B4-605R	14/04/94	28/06/94	
0733	B-2315	China Northern Airlines	B4-622R	25/04/94	12/07/94	
0734	B-2316	China Northern Airlines	B4-622R	27/05/94	27/07/94	
0735	N652FE	Federal Express	F4-605R	25/03/94	20/07/94	*Rachel Patricia*
0736	N653FE	Federal Express	F4-605R	10/05/94	28/07/94	*Samantha Massey*
0737	JA8573	Japan Air System	B4-622R	16/09/94	29/09/94	
0738	N654FE	Federal Express	F4-605R	29/06/94	28/09/94	*Katherine Warner*
0739	B-2323	China Northern Airlines	B4-622R	21/06/94	18/10/94	
0740	JA8574	Japan Air System	B4-622R	12/07/94	30/11/94	
0741	B-2317	China Northwest Airlines	B4-605R	31/08/94	17/11/94	
0742	N655FE	Federal Express	F4-605R	15/09/94	21/11/94	*Don*
0743	N8881	China Airlines	B4-622R	22/09/94	13/01/95	
0744	TU-TAH	Air Afrique	B4-605R	06/10/94	05/04/95	
0745	N656FE	Federal Express	F4-605R	09/11/94	21/02/95	*Devin*
0746	B2325	China Eastern Airlines	B4-605R	23/11/94	28/03/95	
0747	A6-EKO	Emirates Airline	B4-605R	02/12/94	06/03/95	
0748	N657FE	Federal Express	F4-605R	03/01/95	02/03/95	*Lizzie*
0749	TU-TAI	Air Afrique	B4-605R	04/01/95	23/05/95	
0750	B-2327	China Northern Airlines	B4-622R	15/02/95	05/05/95	
0751	F-GSTB	Airbus Inter Transport	B4-608ST	26/03/96	/03/96	
0752	N658FE	Federal Express	F4-605R	30/01/95	04/05/95	*Rudy*
0753	JA8657	Japan Air System	B4-622R	18/05/95	15/05/96	
0754	B2326	China Eastern Airlines	B4-605R	05/05/95	12/07/95	

C/N	Reg'n	Owner/operator	Model	First Flight	Delivery	Remarks
0755	9K-	Kuwait Airways	C4-605R	not taken up	aircraft not built	
0756	HL7580	Korean Air	B4-622R	06/09/95	22/11/95	
0757	N659FE	Federal Express	F4-605R	23/05/95	31/07/95	Calvin
0758	9K-	Kuwait Airways	C4-605R	not taken up	aircraft not built	
0759	N660FE	Federal Express	F4-605R	04/07/95	20/09/95	Michaela
0760	N661FE	Federal Express	F4-605R	24/08/95	25/10/95	Whitney
0761	N662FE	Federal Express	F4-605R	02/08/96	23/09/96	Tessa
0762	HL7581	Korean Air	B4-622R	19/10/95	25/01/96	
0763	B-2330	China Northwest Airlines	B4-605R	23/11/95	08/02/96	
0764	D-AIAW	Lufthansa	B4-605R	18/04/96	08/07/96	
0765	F-GSTC	Airbus Inter Transport	B4-608ST	21/04/97	/05/97	
0766	N663FE	Federal Express	F4-605R	14/09/95	29/11/95	Domenick
0767	B-18501	China Airlines	B4-622R	15/10/96	10/12/96	
0768	N664FE	Federal Express	F4-605R	16/11/95	20/02/96	Amanda
0769	N665FE	Federal Express	F4-605R	21/12/95	19/03/96	Ethan
0770	JA8659	Japan Air System	B4-622R	14/03/96	07/06/96	
0771	N667FE	Federal Express	F4-605R	28/08/96	28/08/96	Sean
0772	N668FE	Federal Express	F4-605R	13/03/96	24/05/96	Tianna
0773	D-AIAX	Lufthansa	B4-605R	23/11/96	30/12/96	Fürth
0774	N669FE	Federal Express	F4-605R	11/04/96	25/06/96	Kaitlyn
0775	B-18502	China Airlines	B4-622R	28/10/97	08/12/97	
0776	F-GSTD	Airbus Inter Transport	B4-608ST	09/06/98	30/06/98	
0777	N670FE	Federal Express	F4-605R	11/04/97	09/06/97	Amrit
0778	N671FE	Federal Express	F4-605R	06/05/97	06/07/97	Drew
0779	N672FE	Federal Express	F4-605R	05/06/97	19/08/97	
0780	N673FE	Federal Express	F4-605R	10/07/97	28/08/97	
0781	N674FE	Federal Express	F4-605R	18/09/97	18/09/97	Thea
0782	HS-TAT	Thai International	B4-622R		02/12/98	
0783	JA011D	Japan Air System	B4-622R		07/05/98	
0784	HS-TAW	Thai International	B4-622R		08/12/98	Suranaree
0785	HS-TAX	Thai International	B4-622R		10/12/98	Srimuang
0786	HS-TAY	Thai International	B4-622R		15/12/98	
0787	HS-TAZ	Thai International	B4-622R		30/11/98	
0788	B-18503	China Airlines	B4-622R		11/09/98	
0789	N675FE	Federal Express	F4-605R	11/05/98	18/06/98	
0790	N676FE	Federal Express	F4-605R		15/07/98	Jade
0791	N677FE	Federal Express	F4-605R	17/07/98	27/08/98	Clifford
0792	N678FE	Federal Express	F4-605R		22/09/98	Allison
0793	N679FE	Federal Express	F4-605R		22/10/98	Ty
0794	N680FE	Federal Express	F4-605R		17/11/98	
0795						
0796	F-	Airbus Inter Transport	B4-608ST			
0797						
0798						
0799	N681FE	Federal Express	F4-605R	30/03/99		
0800	N	Federal Express				
0801	N	Federal Express				
0802	N	Federal Express				
0803	N	Federal Express				
0804	N	Federal Express				

8 CHRONOLOGY

International collaboration and joint development of aircraft was discussed long before the thinking crystallised towards an Airbus. People involved in these exchanges of opinions included Thalau, Valière, Dr Weinhart, Ziegler and Salvador.

24 July 1964 British Aircraft Corporation (BAC) and Sud-Aviation discuss 180-200 seat short/medium-range aircraft.

June 1965 Initial talks begin between the German and French aircraft industry on the occasion of the Paris Air Show. Participants are Dr Weinhardt, L Bölkow and General Puget.

2 July Airbus study group is established in Munich to undertake preparatory work for international co-operation.

6 Aug Deutsche Airbus holds discussions with the Lufthansa executive board on the design of the aircraft, but finds the German flag-carrier unenthusiastic

22 Oct British European Airways (BEA)-led symposium in London brings together Europe's major airlines. Varied reactions to the Airbus concept, but generally negative

22 Dec British transport minister Jenkins visits Bonn to ask the Federal government to participate in a trilateral development of the Airbus. Jenkins agrees to try and persuade France to take a similar line. As a result, Germany's economics minister Schmücker agrees to a 25% participation.

23 Joint-venture team Airbus is founded in Germany tasked with presenting a German project to the Federal Ministry of Economics.

Jan 1966 First negotiations on the design of the Airbus are held by the joint-venture team.

4 Feb Deutsche Airbus visits Fokker in the Netherlands, which considers design too large.

9 March First trilateral meeting about technical design. Britain and France want a twin-engined aircraft for 180 passengers, Germany prefers 250-300 passengers and four engines.

23 Government meeting in Munich fails to agree on technical design. Market analysis is considered wrong.

11 May Similar government meeting in London also ends without agreement, but concession on number of passengers made in follow-up meeting nine days later.

14 June Another London meeting, Britain suggests that its participation depends on the selection of the Rolls-Royce RB.207 power plant.

8 Sept First official consultation between the companies in Germany, France and Great Britain at Hawker Siddeley, at the Farnborough Air Show. Britain and France want the aircraft industry in the three countries to take the lead, as almost a year had been lost through protracted negotiations at government level. This lays the foundations for all future industry meetings.

21 Sept Top level industry meeting in Paris discusses the preparation of joint proposals to be submitted to the three governments, giving evidence of the viability of the Airbus project.

15 Oct Joint proposals and applications for financial participation are presented to the three governments.

6 March 1967 Government meeting in London decision is made in favour of Rolls-Royce RB207-3 engines.

9 May After several postponements of a decision, a trilateral ministerial meeting in Paris approves the undertaking. The industry is requested to supply a joint study by 30 June for an Airbus with two RB207 engines.

30 June Study documents requested at the ministerial meeting are handed over

25 July Trilateral government decision is made regarding the start of the definition phase

Sept Airbus contract is signed by the governments of Germany, France and Great Britain.

4 Sept Deutsche Airbus grouping is registered in Munich, comprising Dornier, Hamburger Flugzeugbau (HFB), Messerschmitt-Werke Flugzeug-Union-Süd, Siebelwerke and Vereinigte Flugtechnische Werke (VFW). Each company contributes DM1 million to the DM5 million registered capital.

Aug 1968 Final decision is postponed until November.

11 Dec Decision is made to build a smaller version, first designated A250, then A300B.

Feb 1969 Federal government decides to continue its support for the Airbus project, even if Britain should withdraw.

March French government announces that Airbus would be built under any circumstances.

10 April British government withdraws amid doubts over the viability of the project.

29 May Airbus development contract is signed by German economics Minister Karl Schiller and French Transport Minister M Jean Chamont at the Paris Air Show at Le Bourget.

RIGHT: The A300 is nearing the end of its production life, but Airbus is flying into a new dawn.

RIGHT: TransAer A300.

18 Dec 1970 Establishment of Airbus Industrie, a Groupe d'Interets Economique (GIE) in Paris.

28 Fokker-VFW signs up with Airbus Industrie, but does not become a full member.

Jan 1971 Agreement for participation of Snecma (France) and Motoren-und Turbinen-Union (MTU) of Germany in manufacture and supply of General Electric CF6-50 engines.

27 May-6 June Mock-up of the A300 is presented at the 29th Paris Air Show at Le Bourget.

30 June Airbus Industrie places contract with General Electric for 20 CF6-50A turbofan engines.

21 Sept CF6-50 engine makes first flight aboard a specially-modified Boeing B52 aircraft

9 Nov Air France becomes first Airbus customer, with a firm contract for six A300B2, plus 10 options.

19 First fuselage assembly completed.

23 First Airbus wing flown by Guppy to Toulouse.

12 Dec Loan agreement for financing initial production is signed with European Investment Bank (EIB) at Luxembourg.

21 Contract is signed by the Banque Européenne d'Investissements and Aerospatiale for the investment necessary to finance the construction of the prototype Airbus A300B.

23 Foreign minister of Spain and the ambassadors of France and Germany sign acccord for Spain's participation on the A300 programme.

14 Jan 1972 Spanish flag-carrier Iberia agrees to buy four A300B4 and take one option.

2 April First CF6-50A engine delivered to Toulouse.

4 A300 is presented fitted with General Electric CF6-50A turbofan engines.

4 May Airbus signs accord with Sterling Airways of Denmark for three A300B4. Aircraft never delivered.

13 July Airbus simulator, built by LMT, is flown to Toulouse aboard a Super Guppy.

28 Sept First Airbus A300 prototype is rolled out at Toulouse, in a joint presentation with the pre-production Anglo-French Concorde supersonic airliner.

28 Oct Airbus A300B1 takes off on its maiden flight from Toulouse, one month ahead of schedule.

19 Dec Lufthansa announces decision to acquire three A300B2s and four options.

29 Airbus completes 100 flight hours in 28 test flights.

5 Feb 1973 The second A300B1 makes first flight.

7 May Lufthansa signs contract for aircraft announced in December 1972.

28 June The first A300B2, and third Airbus, takes to the air.

9 July Swiss charter airline SATA issues letter of intent for one A300B4 for service entry in July 1974 (never delivered).

15 Sept-18 Oct Airbus prototype is taken on demonstration tour to North and South America.

22 Oct Structural static tests are completed.

28 First three Airbus aircraft complete 1,000 hours on 400 test flights.

30 Oct-8 Nov Airbus is demonstrated at Athens, Tehran, Amsterdam, Belgrade and throughout India.

8 Nov-1 Dec Demonstration tour of Africa includes hot-and-high trials at Windhoek, and presentation in South Africa and Zaire.

20 Nov Fourth and last development aircraft, the first production A300B2, takes off on its first flight.

Dec Transbrasil signs letter of intent (LoI) for two A300B2 (never delivered).

5-14 Prototype Airbus is presented in North Africa.

7-16 Airbus No.2 undergoes cold weather trials in Helsinki, Finland.

31 Jan 1974 Final A300B2 certification flight.

15 March Airbus A300B2 receives its certification from the French and German authorities. Testing continues towards certification of Cat III automatic landing system.

10 May-26 June Prototype undertakes demonstration tour of the Far East, also taking in Australia and New Zealand.

23 May Air France operates first Airbus service with a flight from Paris to London.

30 Airbus receives US certification from the Federal Aviation Administration (FAA).

31 July Air Siam places contract for two A300B2.

5 Sept Korean Air Lines signs four firm and two conditional orders for the A300B4.

30 Airbus receives Cat IIIa autoland certification.

22 Nov Charter carrier Trans European Airways orders the second Airbus built, an A300B1, plus one option.

26 Dec Increased-range A300B4 (aircraft No.9) makes its maiden flight.

Feb 1975 Bernard Lathiere is nominated president and chief executive of Airbus Industrie.

11 German carrier Germanair places firm order for one A300 B4, plus an option for another, which is converted into a firm order on 22 July

15 March First A300 B4 demonstrates range-capability with full payload, flying Geneva–Kuwait–Toulouse, a total distance of 9,595km.

26 A300B4 receives certification.

1 April Bernard Lathière becomes the second Airbus president, taking over from Henri Ziegler.

24 Indian Airlines places order for six A300 B2, including three firm and three options.

May First A300B4 delivered to Germanair.

20 June Maximum take-off weight of A300B2 is increased to 142 tonnes.

4 Sept South African Airways becomes first African customer with a contract for four A300B2K and four options. The A300B2K model is optimised for hot-and-high performance with Krüger flaps.

16 Oct Trans European Airways orders one A300B4.

24 Dec French domestic carrier Air Inter becomes tenth Airbus customer with an order for three A300B2.

24 March 1976 A300B4 authorised for take-off at 153 tonnes.

1 April Lufthansa inaugurates Airbus operations.

5 May Transavia leases one A300B4 and orders one more, but order is cancelled in October. Leased aircraft is returned 15 January 1977.

10 June A300B4 obtains FAA certification. Maximum take-off weight is increased to 157.5 tonnes.

27 Air France A300B4 F-BVGG is high-jacked after take-off from Athens on Tel Aviv–Paris return flight, and flown to Entebbe, Uganda via Tripoli. Returned 22 July.

10 Nov First Indian Airlines A300B2 is handed over.

18 Air Inter makes first Airbus flight, carrying government officials from Paris to Nice.

15 April 1977 Thai Airways orders two A300B4 plus two options.

2 May Eastern Air Lines leases four A300B4-2C for six months.

9 Hapag-Lloyd orders one A300B4 plus one option.

6 June Scandinavian Airlines System issues letter of intent for 12 A300, which includes three options

8 Airbus Industrie and Pratt & Whitney sign agreement to equip the A300 with the JT9D-59A turbofan engine.

15 July Aerocondor signs letter of intent for two A300B4.

29 Sept The 100,000 flight hours mark is reached for the A300 fleet. 34 aircraft have now been delivered to 10 airlines.

Dec A300B2 and A300B4 are certificated with the 230.5kN (51,800lb) General Electric CF6-50C1 engine.

9 Aerocondor firms up contract for one A300B4 and one option

30 SAS converts LoI into firm order for two A300B2, and takes 10 options

10 Jan 1978 Airbus launches the 165 tonnes MTOW A300B4-200 model with order from Air France.

4 March Iran Air becomes new Airbus customer with six firm orders and three options for the A300B2.

10 May A300 flies with short-nozzle 233.6kN (52,500lb) thrust General Electric CF6-50C2 turbofan engine.

10 June Olympic Airways places its first order for five A300B4-100, including three firm orders and two options.

26 Airbus wins first order from a US airline, with Eastern Air Lines signing up for 19 A300B4-100 plus nine options.

6 July Swissair signs memorandum of understanding (MoU) for two A300B10 (later redesignated as A310).

16 Contract from Pakistan International Airlines for four A300B4-200 plus six options.

31 First order for the convertible A300C4 from German charter airline Hapag-Lloyd. An A300B4 is also ordered.

18 Aug The Airbus Industrie partners and British Aerospace initial agreement for BAe's entry as a full partner.

30 Malaysian Airlines System (MAS) signs contract for four A300B4-200, including three firm orders.

1 Sept Taiwan's flag-carrier China Airlines orders four A300, plus four options, but contract is cancelled mid-October when the airline fails to receive government approval.

20 Airbus A300 passes 200,000 flight hours. On that date, 54 aircraft are operated by 13 carriers.

15 Nov Philippine Airlines becomes the latest Asian carrier to order the Airbus, contracting to buy four A300B4-100, made up of two firm and two options

28 Alitalia orders 11 A300B4-200, of which eight are firm contracts.

29 BAe's full partnership agreement receives official signature. The British company takes 20% stake in Airbus Industrie with effect from 1 January 1979.

28 Dec Spanish flag-carrier Iberia places firm order for four A300B4-200, plus five options.
Firm orders placed in 1978, including conversions of options, total 70 aircraft.

24 Feb 1979 Garuda Indonesian Airways contracts to buy six A300B4-220 and takes options on six more.

31 March West African multi-national carrier Air Afrique places order for one A300B4-200, together with two A310s.

10 April British low-fare charter carrier Laker Airways orders 10 A300B4-200.

27 Long-established Brazilian airline Cruzeiro do Sul signs contract for two A300B4-200 and takes two options.

29 Egyptair becomes another new customer, contracting for three A300B4-200. Also takes four options.

11 May Singapore Airlines joins the growing list of flag-carriers ordering the Airbus, signing up for 12 A300B4-200, half of which are optioned.

21 Japan's third-largest airline, Toa Domestic places firm order for six A300B2-200.

11 June Agreement is reached between Airbus Industrie and Rolls-Royce.

7 Dec Domestic carrier Trans-Australia Airlines (TAA) orders four A300B4-100, plus two options, for its high-density trunk routes.

5 Jan 1980 Tunis Air signs firm contract for one A300B4-100 and takes option for one more.

30 July China Airlines finally places order for four A300B4, plus one option.

3 Oct VASP becomes second Brazilian airline to order the Airbus, signing for three A300B2.

31 Brazil's long-haul carrier Varig orders two A300C2. Aircraft are assigned to Cruzeiro.

16 Dec Saudi Arabian Airlines (Saudia) becomes first customer for the improved A300B4-600 with an order for 11 Pratt & Whitney-powered aircraft.

6 Feb 1981 Thai Airways becomes launch customer for General Electric CF6-80C2-powered A300B4-600.

30 June Kuwait Airways places first order for the convertible passenger/cargo A300C4-600.

15 Sept Libyan Arab Airlines orders four A300B4-200 and two A300C4-200.

9 Jan 1982 Air Jamaica places orders for two A300B4.

8 July 1983 Maiden flight of the advanced A300B4-600.

9 March 1984 A300B4-600 with Pratt & Whitney JT9D-7R4H1 turbofan engines obtains certification.

26 First A300B4-600 delivered to launch customer Saudia.

3 April Abu Dhabi government signs order for two A300-600.

21 Dec Order is placed by financial institutions for four A300B4 for Pan American World Airways.

20 March 1985 Improved A300B4-600 with General Electric CF6-80C2 engines makes first flight.

26 A300B4-600 receives French certification for Cat.IIIb operation.

1 April Jean Pierson takes over as Managing Director from Roger Beteille who retired.

Sept Continental Airlines places firm order for three A300B4-203.

17 General Electric-powered A300B4-600 is certificated.

26 Launch customer Thai Airways International receives first A300B4-600 with GE engines.

2 March 1987 American Airlines launches the extended-range A300B4-600, initially designated A300B4-600ER and later simplified to A300B4-600R, with an order for 25 aircraft.

9 Dec Maiden flight of extended-range A300B4-600R.

10 March 1988 A300-600R is certified by the French and German authorities.

11 Korean Air orders first A300B4-600R with Pratt & Whitney PW4000 engines.

28 US FAA grants type approval to the A300-600R.

31 Mexican charter Latur orders one A300B4-600.

21 April 1988 Launch customer American Airlines takes delivery of first A300B4-600R

10 May A300-600R enters service with American Airlines.

8 July CAAC orders three A300B4-600R for China Eastern Airlines.

3 Oct Airbus Industrie chairman Dr Franz-Josef Strauss dies.

17 Nov Dr Hans Friderichs is elected chairman of the Airbus Industrie Supervisory Board.

29 Korean Air takes delivery of first PW4000-powered A300-600R.

30 March 1989 Japan Air System places order for 12 A300-622R and takes five options.

3 July 1991 Federal Express places massive launch order for 25 A300F4-600R pure freighter aircraft.

22 Aug Airbus Industrie announces the launch of a heavily modified A300-600 Super Transporter as a replacement for its Super Guppy aircraft — known as the A300B4-600ST Beluga.

20 Oct Special Aircraft Transport International Company (SATIC) is formed as a joint venture between Aerospatiale and Deutsche Airbus to handle programme management of the Super Transporter.

7 Feb 1992 Airbus Industrie places first order for the Beluga.

April 1993 China Northern Airlines and China Northwest Airlines announce orders for six A300-600R each.

2 Dec 1993 All-freighter A300F4-600R lifts off from Toulouse on its maiden flight.

April 1994 A300F4-600R freighter receives certification.

1 Daimler-Benz Chairman Edzard Reuter replaces Dr Hans Friderichs as Chairman of the Supervisory Board.

27 First A300F4-600R freighter is delivered to FedEx.

23 June Airbus rolls out the first A300B4-600ST.

13 Sept A300B4-600ST Beluga takes off on its maiden flight.

29 Sept 1995 A300B4-600ST Beluga obtains its certificate of airworthiness from the European authorities.

25 Oct Airbus Industries receives its first Beluga.

Jan 1996 Beluga enters service with Airbus.

13 Jan 1997 The Airbus partners sign a memorandum of understanding (MoU) to restructure Airbus Industrie into a limited liability company by 1999.

27 June Former President and Chief Operating Officer Bernard Lathière dies suddenly at the age of 68.

July First A300B4 freighter conversion is delivered to launch customer Channel Express.

14 British Aerospace Aviation Services' A300B4 freighter conversion obtains its supplemental type certificate (STC) from the US FAA.

1 April 1998 Noel Forgeard takes over as Managing Director from Jean Pierson, and Manfred Bischoff succeeds Edzard Reuter as Chairman of the Supervisory Board.

23 July An era ends with the death of Airbus Industrie's first President Henri Ziegler.

INDEX